Writing on Air

Terra Nova Books aim to show how environmental issues have cultural and artistic components, in addition to the scientific and political. Combining essays, reportage, fiction, art, and poetry, Terra Nova Books reveal the complex and paradoxical ways the natural and the human are continually redefining each other.

Other Terra Nova Books:

Writing on Water

The New Earth Reader

The World and the Wild

The Book of Music and Nature

Terra Nova
New Jersey Institute of Technology
Newark, NJ 07102
973-642-4673
terranova@njit.edu
www.terranovabooks.org

Writing on Air

edited by David Rothenberg and Wandee J. Pryor

A Terra Nova Book

The MIT Press
Cambridge, Massachusetts
London, England

This book was set in Berkeley Old Style Book by Graphic Composition, Inc. in QuarkXPress and was printed and bound in the United States of America.

Library of Congress Cataloging-in-Publication Data

Writing on air / edited by David Rothenberg and Wandee J. Pryor.
    p. cm.
    "A Terra nova book"
    ISBN 0-262-18230-0 (hc. : alk. paper)
    1. Air—Literary collections. 2. Air. I. Rothenberg, David, 1962– II. Pryor, Wandee J.

PN6071.A56 W75 2003
808.8'036—dc21
                                                    2002038095

10 9 8 7 6 5 4 3 2 1

Contents

List of Illustrations   ix

Introduction   xi
David Rothenberg and Wandee J. Pryor

I   *The Sphere of the Sky*

The Psychedelics of Pollution   3
Harold Fromm

To Build an Imperfect Cloud   11
David Rothenberg

The Birds Keep Their Secrets   20
Howard Mansfield

Lenticular Clouds   32
Harold Humes

Of Aerial Plankton and Aeolian Zones   39
David Lukas

The Bering Bridge   44
Roald Hoffmann

Aerial Imagination: The Character of Air in the Spaces of Le Corbusier and Alvar Aalto   47
Sarah Menin

What Is Seen from the Air   63
Virgil Suárez

Extinct Songbirds of Maine   67
Stephen Petroff

II   *Ineffable Air*

Inebriate of Air   83
John Olson

Clouding   95
John P. O'Grady

From *Oxygen*   105
Carl Djerassi and Roald Hoffmann

The Subtle Humour   114
Bruce F. Murphy

Song of the Andoumboulou: 50   124
Nathaniel Mackey

The Mystery of the Hills   135
Franck André Jamme

From *Of Walking in Ice*   147
Werner Herzog

Thin Air on Mt. Audubon   159
Andrew Schelling

From *The Substance of Forgetting*   163
Kristjana Gunnars

III   *Taking It In, Letting It Out*

Notes on Emphysema   171
Hayden Carruth

The Laugh   181
David Appelbaum

Breath, Air, Voice   189
Andrea Olsen

Acolyte   196
Steve Miles

A Breath of Memory   201
paulo da costa

Aria   211
C. L. Rawlins

The Atomic Nature of Ma   226
Lori Anderson

Surveillance   231
Edie Meidav

IV   *"Blow, Winds, and Crack Your Cheeks!"*

Divide Winds Charge Ultimate Price   245
Reg Saner

There Is No Freedom Spring Air   262
Tõnu Õnnepalu

Holy Wind Before Emergence   265
James Kale McNeley

The Aeolian Harp: An Allegorical Dream   271
F. H. Dalberg

Wind, Play with Me: The Mysterious Aeolian Harp   276
David Rothenberg

The Fallacy of Safe Space   287
David Keller

Wind and Breath   294
D. L. Pughe

Contributors   300

Sources   307

Illustrations

Manuel Acevedo, *#6 Altered Sites,* 1997   **x**

Louise Weinberg, *We Thought There Were Birds,* 2001   **2**

Blur building, ideal image, Diller + Scofidio graphic   **12**

Blur building, simulations of actual conditions, Diller + Scofidio graphic   **18**

Hot air balloon, National Oceanic and Atmospheric Administration photograph   **22**

Lenticular clouds, National Oceanic and Atmospheric Administration photograph   **34**

Marsha Cottrell, *UNT.WT.5.7* (detail)   **38**

Walter Moog, from the book *The World from Above*   **62**

Will Johnson, *Mountaineering Dancers*   **66**

Stuart Allen, *Flying Silk No. 3,* Montezuma Hills, CA, 1995   **82**

Samuel Colman, from *Harper's Weekly,* Oct. 19, 1878   **94**

Joachim d'Alence, frontispiece to *Traittez de barometres, thermometres, et notiometres, ou hygrometres, 1688*   **104**

Danghi Korwa, *Untitled (37)*   **134**

Dhingra Korwa, *Untitled*   **139**

Lahangi Korwa, *Untitled (34)*   **143**

E. M. Bidwell, from *Harper's Weekly,* Jan. 17, 1887   **146**

Ellen Scott, *Self-Portrait in Air*   **162**

Louise Weinberg, *After the Fall,* 1996   **170**

Tuula Närhinen, *Anemographic.* Drawing made by wind and birch, part 1   **180**

Arno Rafael Minkkinen, *Shiprock, New Mexico,* 1997   **188**

Crystal Woodward, *Luberon with Mist, Cherry Blossoms,* 1999   **200**

Tuula Närhinen, *Anemographic.* Drawing made by wind and birch, part 2   **210**

Susan Derges, *Aëris,* 2001   **230**

Jaanika Peerna, *Lines in Silence #7,* 2001   **244**

Tuula Närhinen, *Anemographic.* Light bulb on branches with wind   **264**

Aeolian Harps   **278**

The Dimmitt tornado, National Oceanic and Atmospheric Administration photograph   **286**

Wind rose diagram   **299**

Manuel Acevedo, *#6 Altered Sites,* 1997

Introduction

David Rothenberg and Wandee J. Pryor

As children, we dream of flying, of existing in air without weight or roots anchoring us to the ground. These dreams are foggy nighttime wishes painted on the clouds of our minds. In them we float, drifting above gardens, over houses, mingling with the birds. We play with the wind, allowing it to tickle our skin and tousle our hair. Is it the same fantasy we create in the daytime as we pump our little legs back and forth trying to get the swing to go higher, up into the sky? Once we've reached the limit—a good six feet off the ground—where the chain jolts, we let go, and for a brief moment we feel what it's like to live in air, free from the earth's pull.

Since the beginning of time, people have tried to master air. The captains of the sea rein it in with their sails. The pilots of the sky cut through it with their wings. We have machines to pump air into our lungs, to breathe for us if necessary, and computers to anticipate the movement of the wind. Since "God breathed into our nostrils the breath of life, and man became a living soul," we have been trying to slap a label on air. It makes sense when you think about our need to rule, to define the abstract, to have certainty in an ever-changing world. Unfortunately, that's one of the things air refuses.

Still, we make the attempt. We want to tell you about air. We want to point it out to you, but we cannot touch it or feel it. We take it for granted, just as you do, because we need it in order to breathe or stand up. Sometimes it is a stiff wind on our face, sometimes the stale smell in a closed compartment, or that acrid burn by the great disaster that they tell us is safe but we immediately sense it is not.

The sky is blue, but the color is so far in the distance that we cannot tell where the clear ends and the color begins. The pine needles move in the breeze, but otherwise the air is as invisible, as always. That's the hard part about it, you know. The last book we did was on water, the wet stuff that we feel and require; that one was easy. Everyone has something to say about it—some story, some poem, some memory—whereas air, when we asked for contributions people replied, "Never really thought about it."

But you breathe, don't you, and you enjoy the slap of a stiff wind and fear the power of a tornado. And don't you cringe when you see the color of that brown haze at the horizon, the symbol of human excess, the recurring reminder of pollution's mucky face, and don't you wish it wasn't there?

This book has been harder because air is so ineffable, so flighty, so empty and impossible to catch. It's a bit crazy spending two years seeking contributions on this subject and no other, asking all writers to come up with something on clouds, gales, windharps, and particles of life coursing quietly through the sky.

What most excites us about what we have assembled here is the possible concordance of an airiness in writing itself—an inspiring lightness of being, of words that float through space, of ways of walking and loving that seem to drift gently and elegantly from place to place, never heavy or held down. It is the pieces that present this perspective that might perplex some readers enough for them to ask: Why that? What does *that* have to do with air? while others may alight upon them and exhale with an, "Ah, yes, now I know why the book's put together this way."

Take the selection from Werner Herzog's *Of Walking in Ice*. The great German film director has decided to walk across Europe to visit an ailing friend. He goes straight through the countryside, caring nothing for roads. He looks, he listens, he breathes in the weather and clears out his mind. It is a surprisingly light-filled account for one whose movies are so often dark and heavy. He ends by getting ready to leave the ground and lift off into the air, having crossed the heart of Europe by foot, directly, in the tradition of such literary wanderers as Büchner's Lenz and the nameless travelers in the books of W. G. Sebald.

Kristjana Gunnar's excerpt from *The Substance of Forgetting* is a lovely depiction of a woman seeking definition in the unpredictable landscape of British Columbia's Okanagan Valley. The piece seeks the space in between words, depicting love as a substance without dialogue, an obscure material linking together two distinct forms. The female protagonist of Gunnar's tale is constantly questioning herself, trying to pinpoint where she stands with others and the form she takes alone. Gunnar recognizes that everything that passes between two people passes through space. Like the weather above—under a blanket of fog or wind or rain—the air between us is consistently moving.

Stephen Petroff's "Extinct Songbirds of Maine" is our one nod to those beautiful avian inhabitants of the open sky, those creatures we cannot help but be jealous of as we look up and know others are freer than we can ever be. Petroff's birds are concoctions, dream creatures, metaphors expanded out of history. They too are lighter than the real thing.

We have musical instruments that play themselves in the wind, buildings that masquerade as clouds, and the moment David saw the remarkable drawings by the Korwa people of India's Madhya Pradesh hills in "The Mystery of the Hills," he knew they encapsulated the elusive aesthetic he envisioned for this book. Why? The Korwa are a people with no written language, but they enjoy playing with language. Taking pen to paper, they explore shapes and forms similar to Hindi and Sanskrit, without knowing exactly what they mean. It is as if they are spontaneously inventing language through the air, from the mind and onto the page. The images are fresh, abstract, yet rooted in detached yearning for the literate, the civilized. They like to draw, even though they do not know how to read or write. Their art marks the waft of language like wind on paper. They are not held back by the restrictions and rules of the written word. They are the people who write like the air, suddenly, elegantly, and with inspiration, not regulation. Compare their drawings with Tuula Närhinen's pictures made by brushes attached to tree branches, moving in the wind, and you will see.

This book wouldn't be complete if we discussed only the joy of breath, the experience of laughing or singing, without paying tribute to those ailments and afflictions that restrict our ability to breath. Harold Fromm's essay, which paints a vivid image of pollution, will have you cataloguing your headaches and dietary habits, questioning if it's the dark smog overhead that's causing you to feel sleepy by midafternoon. We've also included Hayden Carruth's prose poem on emphysema in which he explores a reality that is incomprehensible for most of us—a state where you are conscious of very breath and grateful for an inhale lacking pain.

We have bits on hot air balloons, clouds, swirling dervishes in the sky. We've attempted to address air's rich, diverse history, a past that vacillates between science and spirituality, from aerial plankton to Navajo wind gods. In an excerpt from *Oxygen*, you'll see air's initiation into science as Hoffmann and Djerassi show the founders of "vital air"—Priestley, Scheele, and Lavoisier—fighting over who can claim its discovery.

And do not think the elements stay easily apart; they mix and mingle, whether or not they exist. Don't expect any more element books from us either. We are going to take a break. The next book will have something to do with progress, with change, as it moves through the realms of the human and the natural—our usual obsessions—a subject that covers just about everything.

This book is eclectic, to say the least. Broken down into parts, it looks at air as substance and metaphor, examining weather and objects in the sky. We have sought to give it back its elusiveness. Air escapes us, but it also pervades everything we do, from the laughter that trickles out of our mouths to the light it carries shining down onto this page. It unites us together, erasing contradictions by being everything at once. It gives and it takes; it allows movement and forbids it. Air can create glorious sights, radiant skylines, dancing trees while at the same time it ravages and destroys whole towns. The air we take in is the same air that pulls boats up to shore, empties trees of their branches, sings on the telephone lines.

So air is here subject but also style. When you have perused the book and found what definition of air we present by example, we hope you will feel the breath of the world flowing in and out.

I

The Sphere of the Sky

Louise Weinberg, *We Thought There Were Birds,* 2001

The Psychedelics of Pollution

Harold Fromm

Surely, nothing is more reproachful to a being endowed with reason, than to resign its powers to the influence of the air, and live in dependence on the weather and the wind for the only blessings which nature has put into our power, tranquility and benevolence. This distinction of seasons is produced only by imagination operating on luxury. To temperance, every day is bright; and every hour is propitious to diligence. He that shall resolutely excite his faculties, or exert his virtues, will soon make himself superiour to the seasons; and may set at defiance the morning mist and the evening damp, the blasts of the east, and the clouds of the south.
—Samuel Johnson

He had, till very near his death, a contempt for the notion that the weather affects the human frame. . . . Alas! it is too certain, that where the frame has delicate fibres, and there is a fine sensibility, such influences of the air are irresistible. He might as well have bid defiance to the ague, the palsy, and all other bodily disorders. Such boasting of the mind is false elevation.
—James Boswell

Alas indeed! The influences of the air are even more irresistible than Boswell's prescience could have envisioned. After Darwin, Marx, and Freud, the arena of human freedom has come to seem painfully shrunken. And after contemporary environmental studies, even less remains. But recognition of environmental constraints upon our behavior can at least inform our options, as we come to see how many "choices" are actually made for us by the nature of things.

My own knowledge of these matters springs from the personal experience of having lived in the dramatically polluted environment of northwest Indiana, not far from the steel mills and power plants that line the shores of Lake Michigan from Chicago to Michigan City. And since the essence of that knowledge introduces what might be called an existential environmentalism, its accumulation from first-hand experience lies at the heart of the whole affair. For despite the almost unbelievable development of knowledge about environmental issues during the past two decades, the personal realities of the problem of bad air remain almost unexplored. The dependence on technological means of measuring air quality has presented a very skewed picture of what it means—what it feels like—to be a person amidst pollution and this picture continues to leave the erroneous impression that pollution is a somewhat abstract problem that affects other people rather than me myself or that at bottom it is mainly an esthetic nuisance. Considering the limited extent to which most people are aware of the ways in which pollution is concretely affecting them now at this very moment, public support of environmentalism exhibits a not-so-common instance of disinterested human concern. We have repeatedly been told that polluted air is bad—for plants, for animals, and for other people—and we have responded accordingly. But how, we need now to ask, is pollution bad for me—as I take this very breath, as I try to go about my daily tasks with the sense, probably illusory, that I am in control; now, as I am reading these words, not merely in some nebulous future when I may learn that I have become a casualty?

The media, spurred on by government agencies, have helped to foster this unbalanced picture of pollution as bad for others. The constant warning is that cardiac and respiratory patients should stay indoors today (in their private air supplies?), or possibly joggers along Lake Michigan should take it easy because of the dangers of Ozone Alley, the lakefront from Waukegan to south Chicago. But presumably, the rest of us are fine. Television weather reports rarely mention pollution, except in places like Los Angeles, even though there is literally no such thing as weather separable from the vast trans-continental air currents that conduct emissions hundreds, even thousands, of miles. And when they do mention it, they mislead us by reporting that air quality in Chicago was fine today, while ignoring the fact that pollutants had blown out into the suburbs or countryside, to whose residents it was far from fine, even though they have been told otherwise. The so-called urban and suburban "haze," which sounds so harmless and romantic; the snow squalls that plague southwestern Michigan and northwestern

Indiana even though it is snowing nowhere else, the stench of sulfuric acid emanating from suburban snowbanks—these are not simply "weather." For Commonwealth Edison and U.S. Steel produce as much of it as God.

To further confuse our understanding, the data upon which reports are made about health problems are gathered mainly from hospital records, that is, from the most extreme and dramatic cases. This helps to explain why there is so much misleading emphasis placed upon cardiac and respiratory patients: these are the cases that end up in hospitals and thus are the most visible. But the effects of air upon the general population are rarely discussed—and rarely will be unless reports are made about people who are not in hospitals. Given the nature of technological society, it is very difficult to obtain information about matters that cannot be easily fed into a computer. But ordinary people show every sign of being affected by pollutants as much as "sick" people, even if they don't end up in a hospital. And these maladies occur not just among dwellers in smoky cities, but among people who live in the suburbs, in the country—in a word: everywhere.

In fact, there are few if any "safe" places that are so far from pollution sources as to be exempt from today's intimations of mortality. When my wife and I moved from our rural farm fifteen miles south of Gary, Indiana, to our present suburban home about sixty miles northwest of Gary, we naively believed we would be far enough from the steel mills to escape the air that was making us ill more than half the days of the year. But Chicago's northwest suburbs, superior as they may be to northwest Indiana, are polluted enough. Southeast winds blow Gary's emissions deep into Wisconsin, while Waukegan's power and steel plants send their devastating pollutants to the already overburdened Illinois atmosphere whenever winds are from the north and east. With Joliet sending its own contributions on southwest winds, very few days of clean air are available in the greater Chicago area. And they may be getting fewer as coal becomes the latest panacea.

Up in Madison, Wisconsin, 160 miles or so north of Indiana, one can still discover the plume of bad air from Gary-Chicago if the winds are right, with Milwaukee sending off its own toxic clouds. And if that seems far, on one of our visits to the English West Country, the Cornwall of legend and song, my wife and I began to experience the similar physical and mental symptoms of bad air, with much disbelief and despair, only to learn a few days later from the London *Times* that the Ruhr Valley's heavy industry in faraway Germany is a major source of British air pollution. In the United States, the vines of grape growers in western New York state suffer marked damage because of emissions from Gary; New York City is said to receive much of its bad air from Pittsburgh, Gary, and Birmingham; while New

England inherits it all as acid rain. Indeed, the *New York Times* of Feb. 8, 1979, reported: "Smog and dust from industrial Europe and China may account for a mysterious haze that hangs over Alaska, Greenland, and the Arctic Ocean every spring, according to analysis of atmospheric particles that are assumed to cause haze." Whither then escape? Denver has had its day, Los Angeles and Riverside are cesspools of pollution, Ohio a nightmare, Phoenix in decline, and Waukegan bad for your health. The circles increase in contemporary Hell.

There are a number of ways of recognizing air pollution without specialized instruments. Visually, the most dramatic signs are familiar by now to most people in city or country: orange-brown smog that obscures almost everything within range and which, in its lesser presence, is euphemistically referred to as "haze." Apart from the more dramatic cases—inability to see the Hollywood Hills, amorphous quality of skyscrapers just across a New York street, even in sunlight—smogs and hazes are not so easy to notice unless one lives in the country or atop a high-rise. Distance is required to get an accurate picture, since the sky directly overhead almost always seems to be clear, giving the comforting impression that one's self is almost never amidst the pollution. As one drives into the city the nimbus of smog seems far ahead, hovering over tall buildings, appearing to recede on arrival, until—miraculously—the city does not seem polluted when you are in it! But if there is haze on three sides one must obviously be in it, however clear overhead. Furthermore, some of the most intensely polluted days consist of gaseous emissions with little particulate matter, and only a careful scan of the horizons hints at the presence of sulfur dioxide or ozone in what otherwise appears to be a clear and sunny sky.

Smells, another sign of emissions, are harder to detect because one's sense of smell is dulled very quickly. Often, a sudden exit from the house will reveal stronger concentrations of odors that had infiltrated too gradually to be discernible inside. Odors suggestive of asphalt or tar as well as oil and "baked potatoes" are common by-products of steel mills and petroleum plants and the smell of ozone on summer mornings can be detected in and around urban areas with heavy traffic from automobiles.

But it is the physical and psychological signs of air pollution which are the most important and for which the visual and odorous merely provide confirming evidence of the extent to which one's total being is shaped each day by the particular chemical mix of the air. The range of these symptoms is great and most often they are found in combinations rather than singly. Though many may also be found in connection with maladies unrelated to the air, it is the circumstances of their combination that enable them to be traced to industrial and automobile emissions.

When evidence of sight and smell accompanies these experiences, when the wind is blowing from the direction of major pollution sources, when one's self and friends feel these maladies at the same time, and when the symptoms vanish with a shift in the winds, only sheer perversity can fail to take the hints.

Although coughing and burning eyes are most familiar, and although official warnings usually stress respiratory ailments, these discomforts are only the most obvious and operatic of the effects of bad air. More widespread and insidious than these are a broad variety of headaches, often combined with nausea and dizziness, especially on arisal in the morning. One is likely to have slept through the night as if under sedation, awaking dizzy, drugged, and in a stupor. This commonplace "inability to get up in the morning" is not just a donnée of human life, however, but a gift of industrial society. Once up, one may feel unsteady, heavy-headed, with a growling stomach and heartburn, preoccupied during breakfast and in-attentive later in the day, unable to focus or maintain a clear train of thought. Nor does it matter if one has had 6 or 16 hours of sleep, since the problem is not lack of sleep. If one is suffering from hunger and heartburn, the hunger does not go away, even after a decent breakfast, but rather insistently gnaws, producing an in-satiable craving for high carbohydrate junk foods. On badly polluted days people wonder why they are eating all those potato chips and candy bars when they have had their usual breakfasts, and the junk food machines will be whirring away. Thoughts of dinner start crowding the mind early in the afternoon. In the highest realms of diplomacy, the august diplomats, bowed by the weight of international affairs, can only think of sweet sherry and scones. Reading errors, typing errors, carpentry mismeasurements start to increase, along with an ill-temper that seems to be caused by nothing in particular. The more sedentary the activity, the more one is at the mercy of out-of-focus intellect, though joggers are warned to take it easy.

Like the extreme difficulty getting out of bed in the morning and the curious hunger even after eating, the pseudo-cold is not yet generally associated with badly polluted air, but only a little attention is needed to make the connection: a sore throat suddenly seems to be developing, one's arms and legs, joints and muscles begin to ache and seem sore and fatigued, and one's head feels heavy as in the onset of a cold. One wishes simply to fall into bed and give way to the miseries of a cold. But then next day, magically, all of this disappears. One forgets that a cold was even settling in. The winds have shifted. Where are the colds of yester-day? They have literally blown away. For me the most striking example of this took place a few years ago in London. I had arranged with a friend early in the week to go on a weekend walk in the country, but by midweek I began to feel completely overtaken by cold symptoms and lethargy. When I took to my bed, I phoned to cancel the outing and learned that both my friend and her office co-worker had begun to feel exactly as I did on the same day and felt that way still. During the following week we reported to each other that our "colds" had gone away on the same day, when indeed the persisting winds had shifted direction.

Depending on the industrial mix, one is apt to have nausea and dizziness, pains in the hands and feet, chest pains, heartburn and gas, muscle aches, burning eyes, lethargy and headaches, to mention common signs. But beyond the physical are depression and dispiritedness about nothing in particular, short-temper, irritability, aimlessness, a tendency to quarrel, an inability to read, a general inattentiveness and a despair of the possibility of human happiness. One is conditioned to a frightening degree by the day's particular—and particulate—mix. A midday shift in the wind can change one's philosophy as dramatically as one's cold symptoms.

Reflecting on all of this one is led to ask, Why, after all, should the experience of "just feeling out of sorts today" be exempt from the causality that lies behind all other kinds of phenomena? These experiences must have their causes like any other, however hard to pinpoint. And so, when it finally becomes apparent that transitory mental and physical states are not just causeless random happenings, the discovery of pretty consistent patterns becomes not only practical but inevitable. Because periods of malaise can range from only a few hours to many days in a row, depending on atmospheric conditions, it is rarely possible to have these experiences diagnosed by a physician. The ailment is gone by the time the appointment day has come around; yet even if it were still present it is highly unlikely that the average M.D. would provide a correct diagnosis. For most medical doctors know little or nothing about the psychological or physiological effects of pollution. To further complicate things, the sudden recoveries from environmentally induced ailments cause the sufferer to forget about them as soon as they have gone away. When they recur, the same kind of cycle is apt to take place. One never gets any closer to the solution of these problems.

So one must try to solve them by oneself. Learning to ascertain the correlation between physical/psychological symptoms and air pollution requires an awareness of the principal sources of pollutants for any given section of the country. A representative picture can be derived from the Chicago area: the entire south end of Lake Michigan produces massive quantities of air pollution, much of which is often strikingly visible and smellable while driving on the Chicago area expressways. With southeast winds, to use one limited illustration, a vast cloud of pollutants starts out from a fairly narrow area south of the city (from a triangle that includes East Chicago, Gary, and Michigan City) and spreads in an ever-widening wedge over Chicago and its northwestern suburbs, a wedge that is easy to see while driving on the tollways that circle the city. If the wedge remains narrow, one can drive in and out of it, sometimes more than once during a trip in and around the city. When it is broad, this wedge spreads out well into Wisconsin. But if you happen to be close to its starting point, where it is very narrow, it is actually possible to miss its effects altogether even though literally millions of people are living in its shadow at that same moment farther away.

A drive around the perimeter of the city on the Tri-State Tollway can be an educating experience, for it is possible to move between sunny, clear, and beautiful skies at one end (i.e., the Indiana or the Wisconsin border) and dark, foul smog, or even a very confined snowstorm at the other, all within an hour and a half. If one is alert, it is possible also to observe the physical and mental transformations that may take place as one enters and leaves the different mixes of air. Headaches and dirtiness can appear and disappear as one rounds the large curve from O'Hare airport to Hammond, Indiana, accompanied by the "baked-potato" smell, the Sherwin-Williams smell, the oil refineries smell, and the smell of *real* potatoes from Jay's potato chip factory.

Although the effects of being downwind of a pollution source can be very pronounced, it is only when people compare notes about how they feel that illuminating causal connections can be made. When out of town friends were visiting me on what seemed to be a beautiful, cool, clear summer's day with light northeasterly winds from Waukegan (a dependably bad source of pollutants), they suddenly announced that they both felt so miserable that they would have to nap for a while. It must be air pollution, they told us, because they had headaches, felt drowsy and lethargic and could hardly keep their eyes open. And to top it off, they were in very low spirits, verging on despondency. We had talked about such things with them before, and now had to agree that the air was pretty bad. My wife and I had struggled out of bed that morning and tried to be lively hosts even though we felt lifeless, dazed, and unfocused all during the day. Misery had the company it so often wants, which helped to cheer us all up.

A new dimension can be added to the old philosophic chestnut about free will: it is not merely one's genes, one's prior psychic history, one's parents, social class, etc., that determine one's accomplishments, moods, and perspectives. It's the chemical mix of the very air one is breathing at any given moment, for breathing such air is a counterpart of eating food contaminated with pesticides or drinking water laced with asbestos fibers or PCB's, except that the effects are often very immediate. How "free" is a creature whose worldview at a given moment has literally been concocted miles away in the vat of a steel mill? If one can be drugged without pills, soused without Scotch, depressed without precipitating psychological events, irritable without irritants, and pessimistic without philosophy; if one can be hungry without fasting, exhausted without having expended any energy, and afflicted with heartburn and indigestion without recent food, then what does it mean to have a mind or a will of one's own?

Seen in this light, various kinds of experience shed their metaphysical mists and encourage a sordid behaviorist perspective. For instance, in the *Chicago Tribune* of Jan. 20, 1980, the food writer Carol Haddix made some routine observations about her uncontrollable hunger:

> The hunger pangs were beginning. It usually happens around 10 A.M. whether breakfast is eaten or not. It becomes difficult to work with that nagging stomach rumbling away. Coffee or a cup of tea works to stop those pangs for a short time, but your body knows better. "Where's the food?" it cries.
>
> I usually last until about 10:30 or 11 A.M. before I make a mad dash to the junk-food machine and wolf down a disgusting candy bar or a bag of potato chips. Oh, if my mother only knew.

If only the writer herself had been in a position to know: a polluted morning in Chicago, an insatiable craving for high carbohydrate foods despite breakfast. A check of the day's pollution readings would doubtless reveal more than one's seeming-wise body that supposedly "knows better." Though perhaps it does know better, since air quality is causing it to crave the sort of nutrients that provide quick energy.

When junk food machines are cranking out their chocolates, when professors wonder why their classes are afflicted with lassitude and inattentiveness on Monday, while they are very lively on Tuesday despite less interesting subject matter, when office workers can barely do their typing—when all this is finally observed and tallied up, new knowledge has become possible.

Boswell's intuitive awareness of more than two hundred years ago may now be ready for general circulation.

(1983)

To Build an Imperfect Cloud: Diller + Scofidio's Blur Building

David Rothenberg

All architecture involves the human containment of natural space, so of course it is concerned with air. But how many buildings are themselves made of air? Perhaps only the Blur Building, designed by the firm Diller + Scofidio for the Swiss National Expo of 2002. Situated at the base of Lake Neuchatel in Yverdon-les-Bains, Switzerland, it opened in May 2002. By the time you read this, it may already be closed forever, or else it has become a permanent attraction, although it will be operable only during the warm months or in a warm climate.

Project leader Dirk Hebel, whom I spoke with at the firm's New York offices, tells me it's not really an air project but a water project, perhaps more appropriate for our previous book, *Writing on Water*. The pavilion is made out of filtered lake water shot as a fine mist through 34,000 fog nozzles, creating an artificial cloud that measures 300 feet wide by 200 feet deep by 65 feet high. A built-in weather station controls fog output in response to shifting climatic conditions such as temperature, humidity, wind direction, and wind speed. The public approaches Blur by a ramped bridge. The 400-foot-long ramp deposits visitors onto a large open-air platform at the center of the fog mass, where movement is then unregulated. Visual and acoustical references are erased along the journey into the fog, leaving only an optical white-out and the white noise of pulsing water nozzles.

Terra Nova: When we set out to put together a book on the theme of air, we knew it was going to be tough. Water was so much easier! But when we heard you were constructing a building in the form of a cloud, we knew this was the kind of thing that should be in a book on air.

Blur building, ideal image. Courtesy of Diller + Scofidio

Dirk Hebel: But isn't that funny? From the very beginning, we never actually conceived of it as a cloud or the image of a cloud. We hate the association of this project with the image of a cloud!

TN: Why?

Hebel: Because it puts this image in peoples' heads, and they close down from that. It could be anything. Why a cloud? It could be any project that you want it to be. I even hesitate to call it a building; it's more like an atmosphere. The greatest architectural interest was in creating something that's constantly shifting. The concept works against the well-known definition of architecture, which is static, which always stays the same, which never changes. You can come whenever you want. It's always the same. The idea for the Blur Building was—and I think we actually achieved it—to create something that moves every second, that looks different from one moment to the next.

TN: So what does it mean when you say you're creating an atmosphere?

Hebel: It's nothing else but a space that contains little water droplets, which can be moved by the wind and other weather conditions.

TN: In some conditions, could your creation just drift away?

Hebel: Yes. That will happen with strong winds. You'll see a trail, like fog on a leash.

TN: How often will it be contained? How often do you think it will look like this beautiful, well-rounded picture?

Hebel: Maybe once during the entire summer-long expo. This is a bit of a problem, because people will expect that when they come to see it, even though, for the last two years, we've been saying that it will not look like that. The image is a million times stronger then anything we are able to say about the project.

TN: What does that say about the building if the image is more powerful than its reality?

Hebel: It says that our world is image oriented. Everything is about image. What's now called a pavilion is acting against its tradition, acting against its kind of power and what a pavilion should usually have. This is a pavilion that is inexistence in itself. It contains its inexistence. There is always this kind of ambivalence between those two stages.

TN: So sometimes what's supposed to be inside it is going to be outside. How do you know when you've crossed the line?

Hebel: Say it's a building—of course, it has a structure; it has a circulation system, but it has no skin. It doesn't protect you from certain types of weather conditions. So in some sense, it is a building, and in some sense it's not. For example, the fire department asked us to put a sprinkler system inside. We told them the building itself is a sprinkler system. These are all issues but, like the most pragmatic people, you can't involve them in an architectural discourse.

Usually when you go into a building, you take off your coat, but in this building, you have to put something additional on—a special raincoat, which we will provide. And where it stands right now, people have to choose if they want to buy one or not.

TN: Or they can just be wet.

Hebel: Actually, on a sunny day, a warm day, it will be such a relief to go into the Blur Building. It's really nice. You don't get wet, just a little spray. I think on a windy, rainy day, you would want to put something on.

TN: Well, clearly, if it's a building, there ought to be something inside it—something you have to experience.

Hebel: There is something inside. Imagine all the noise coming out of the nozzles. It's like turning on a gas stove. It's a really great sound. It will be the biggest mist array ever created in the world.

TN: But how is it architecture and not sculpture? What is the difference?

Hebel: It has two qualities to it. If you are on a boat and see it from far away, it has this iconographic value—an image that's very strong—but it also has that shape-shifting reality that negates that image. How do you describe fog? It's something that blurs things, makes them less distinct than they actually are. So when we conceived of the project we called it Blur—the Blurring Building. So it embodies negativism in itself.

TN: Is there stuff in the building?

Hebel: How we hold the whole thing up is that we build a structure on the side, which is a tensegrity structure—a structure that was developed by Buckminster Fuller in the thirties. And what you can do with that tensegrity structure is actually create long spans with steel and tension and pressure members. We have only four pillars actually going into the lake, and along the tensegrity structure we apply all our fog nozzles and fog lines. So there are actually walking bridges into that atmosphere—what we call the fog atmosphere, the water atmosphere. And it has another platform that we call the media platform, which allows people to wander around without any restrictions. They are walking on a metal grid, and the fog is falling around them, above and below the deck.

TN: So you think this project is more about water than it is about air. How does the water become air?

Hebel: It's a kind of technology transfer. From the beginning, we thought the idea was great, and everyone just loved it. Getting into the project more and more, we realized how difficult it actually is. We had a prototype running between spring and summer last year, and during the test period, we ran into many problems. I think your question is right on. "Is the project a water project or an air project because the water helps us to create the fog but the biggest enemy for this project is the air?"

TN: The air is the enemy?

Hebel: Because fog is very sensitive to wind—to wind speed, to wind direction, and very, very sensitive to humidity. For example, a very dry, warm, high wind that comes down from the mountains blowing into the fog structure can influence the fog production in a very negative way. Basically, we do nothing but take lake water, pump it from the lake, filter it, prepare it, put it back into a pump system, and pump it with high pressure into the feeding lines of the fog system. And then all you have is a little nozzle with a pinpoint where the water is pushed with high pressure into a spray.

The difficulty with our project is that we don't treat it as art. We don't want to have it just as a sculpture floating around. We have to care about it. We have to control it. So we have these stations around the structure where we can monitor the weather, and we have computers running the systems. Depending on the weather, the computer will tell the pumps—and we have eighty pumps now—what to do and when. So behind that whole nothingness is a highly technological system.

TN: How can you contain a cloud?

Hebel: You can't. The only thing you can actually do is produce enough fog so that when it comes out, you have enough to cover your structure, but what happens to the fog after you cannot control. Sometimes we have hundreds of meters of fog over the lake, but we like this because it is the atmosphere of the building, constantly changing shape, changing size, changing its physical condition.

TN: Do you think all buildings are about air, especially since they contain space?

Hebel: No, not at all. You have to deal with air in all buildings—the air-conditioning and the circulation of air. Of course, as architects, we have to think about this, but I don't think many architects actually use air as a substance out of which to build.

TN: Well, buildings take up space, which is filled up with air.

Hebel: But usually it's not something that's actually taken on as a philosophy, as a content, in that sort of sense—designed around.

TN: Is there any sort of precedent in architecture where buildings have been formed out of air?

Hebel: There are inflatable structures. They must consider the air, the pressure, the tension of the walls. But this is very contained again. It's nothing you're exposed to.

TN: What do you think it will be like to be inside the Blur Building?

Hebel: Oh, I know what it's like inside. You feel completely disoriented, and you have this really, really nice sound around you of these fog nozzles. It sounds almost like a gas, like when you turn on a gas stove. It's a little sizzling. And you have this nice freshness that you're inhaling.

TN: Is it naturally fresh, or is it made fresh?

Hebel: It's cold, it's filtered, it's clean. This is Switzerland, so the water is very clean to start with. You have this really cold air because the temperature is dropping inside the structure because of the water. What we also found is that when the water hits the needle of the nozzle, a lot of electrons are put into the air, and your brain gets a little kick out of it. It's almost a little high kind of thing. We didn't believe it in the beginning when everyone told us that, but then we went in there, and it really does happen.

TN: Does that happen naturally in fog?

Hebel: It happens, actually, in thunderstorms.

TN: Yes, you get slightly electrified.

Hebel: Yes, yes, exactly. With a single nozzle, it creates so much electricity that you can light a light bulb by running a wire through the space.

TN: So are you going to put light bulbs in there?

Hebel: [Laughs] That would be funny. No, we won't do that. But we have a lighting system inside, for it to run at the night.

TN: Do you want people to feel totally lost inside?

Hebel: Yes, that's one experience. And actually you create it with minimal effort. Of course, people get lost when you overwhelm them with a bunch of information, but that's exactly what we didn't want to do. We want to take away—to erase context instead of adding context.

TN: I see from this model that people will be able to climb to the top of the building.

Hebel: Yes, what we have on top of the structure is what we call the angel deck, so people can have the experience of walking up the stairs to the top of the structure and looking down on the fog.

TN: Will you always be on top of the fog, or is that impossible?

Hebel: You'll actually be on top.

TN: So it can be contained, even if it's streaming off to the sides.

Hebel: And we hope to install a bar that serves only water. You will be able to buy fifty to one hundred kinds of water from all over the world.

TN: You could have an oxygen bar too.

Hebel: [Laughs] There will also be a light show at night.

TN: Really?

Hebel: You can project on clouds.

TN: Will the images be crisp?

Hebel: Of course.

TN: So the image of the building remains indistinct, but you can project precise images onto its flanks?

Hebel: Exactly. It has to offer more images all the time. And there will be more interpretations of it. And that's also why we try to avoid interpreting certain things, because we think any interpretation is right. There's no wrong reading of our work.

TN: How does that play out?

Hebel: It's very hard because the press in Switzerland is focused on that one perfect cloud image.

TN: That image has appeared many places?

Blur building, simulations of actual conditions. Courtesy of Diller + Scofidio

Hebel: Yes, of course.

TN: You don't like it?

Hebel: No, we would rather have different images instead of the same one, but people like to know the icon of an image. I can understand that, but it would be nice if, slowly, other types of images appeared—like the cloud in different weather conditions and showing how those conditions will affect it. It's like what I said: every single second, that building will look completely different.

TN: That singular cloud image of the building is so compelling. And by the time this book comes out, this project is going to be up and down. But we'll be sure to include an image of the building as constantly changing.

I'll try to get there to photograph it myself, to prove that it really existed! Now that I know more, I see the project as kind of a movement between water and air. I can see you don't want to say too much about the work, as if to give it away. And you don't want it to mean anything in particular. But it can't just be a piece of artwork that stands for itself. It's a building, so there has to be an inside and outside, interior experience and exterior experience.

Hebel: It's undefined. That's the beauty of it. It's indefinable. That transition is actually underdefined; it's blurred. The most interesting thing for us is that we challenge the fixity of architecture. You blur in different ways—you blur content, you blur form, and you blur the edge of architecture—and that's something I think is very interesting. The thing that's funny about it is that the whole blurring process takes quite a lot of energy and quite a lot of precision.

The Birds Keep Their Secrets

Howard Mansfield

I

As a child on a coffee plantation in Brazil, Alberto Santos-Dumont played the game of "*passarinho—voa*," "the little bird flies." Sitting around a table, the players would quickly say, Does a pigeon fly? Does a hen fly? and the children would raise a finger. But if someone said, Does a fox fly? and a player's finger was still raised, he lost. Whenever they said, Does man fly? "I would always lift my finger very high, as a sign of absolute conviction. . . . The more they laughed at me, the happier I was, hoping that some day the laugh would be on my side," Santos-Dumont recalled years later in Paris, after he won the Deutsch prize for a guided airship flight. Following a predetermined course, he had steered around the Eiffel Tower and returned to the starting point. No one had done that before his triumph in October 1901. He was a short, thin, serious-looking man who dressed for his flights in a dark suit, high starched collars, tie, and a bowler hat. "Le Petit Santos" was brave, innovative, and stylish. He was celebrated in the City of Light, and honored with a medal back home in Brazil.

"They play the old game now more than ever at home," a friend wrote to him after his flight. "They call it now 'Man flies!'"

When the young Santos-Dumont first went to Paris with his family, he was surprised to learn that ballooning had not advanced since the Montgolfier Brothers had made their ascents a hundred years earlier. There were no "steerable balloons." A balloonist lived on "the chances of the winds."

Santos-Dumont was determined to go aloft, but the aeronauts discouraged him, so he bought an automobile instead, a three-and-a-half horsepower Peugeot roadster, a rarity in 1891. "Such was the curiosity they aroused that I was not allowed to stop in public places like the Place de l'Opéra for fear of attracting multitudes and obstructing traffic," he said. When his family returned to Brazil, Alberto packed his Peugeot for the voyage.

At age eighteen his father gave him his liberty and his inheritance. He sent his son back to Paris, warning him that Paris was "the most dangerous place for a youth" and giving him this advice: "Do not forget that the world's future is in mechanics." Alberto agreed. He had loved to tinker with the plantation's machinery. He had seen his first working petroleum engine at a Paris exposition. "I stood still before the motor just as if I had been nailed down by fate," he said. He had been nurturing a secret desire to build flying machines. "In the long, sun-bathed Brazilian afternoons, when the hum of insects, punctuated by the far-off cry of some bird, lulled me, I would lie in the shade of the veranda and gaze into the fair sky of Brazil, where the birds fly so high and soar with such ease on their greet outstretched wings . . . and you have only to raise your eyes to fall in love with space and freedom. So, musing on the exploration of the aerial ocean, I, too, devised airships and flying machines in my imagination." He kept these thoughts to himself. "In those days, in Brazil, to talk of inventing a flying-machine, or dirigible balloon, would have been to stamp one's self as unbalanced and visionary."

Back in Paris he went aloft. He loved his first balloon ascent, and it seemed as if he "had really been born for aeronautics," he said. "Infinitely gentle is this unfelt movement forward and upward. The illusion is complete: it seems not to be the balloon that moves, but the earth that sinks down and away." Two miles up, the sun cast the balloon's shadows on the "dazzling white clouds" below. "Our own profiles, magnified to giant size, appeared in the center of a triple rainbow!" Alberto had packed a "substantial lunch" for himself and the two aeronauts, consisting of "hard-boiled eggs, cold roast beef and chicken, cheese, ice-cream, fruits, cakes, champagne, coffee and chartreuse." "No dining-room can be so marvelous," he said. "A joyous peal of bells mounted up to us. It was the noonday Angelus, ringing from some village belfry." He finished his little glass of liqueur as they set down. He was hooked. "I cannot describe delight, the wonder and intoxication of this free diagonal movement onward and upward. . . . The birds have this sensation when they spread their wings and go tobogganing in curves and spirals through the sky."

Santos-Dumont was free. He ballooned around Europe. He flew at night, rising "through the black solitudes of the clouds into a soul-lifting burst of splendid

Hot air balloon. Courtesy of the NOAA Collection

starlight. There, alone with the constellations, we await the dawn! And when the dawn comes, red and gold and purple in its glory, one is almost loath to seek the earth again, although the novelty of landing in who knows what part of Europe affords still another unique pleasure." He had crashed and flown in storms. Lost alone at night in a thunderstorm, "I felt myself in great danger, yet the danger was not tangible. With it there was a fierce kind of joy. What shall I say? How shall I describe it? Up there in the black solitude, amid the lightning flashes and the thunderclaps, I was a part of the storm!"

Earlier attempts to invent a steerable balloon had failed. Electric motors and steam engines had been tried, but they were no match for the "vindictive winds," as the balloonists said. Alberto adopted "so uncertain an engine as the petroleum motor of the year 1900." His experimental method was direct: he had himself hoisted in his tricycle-automobile up into a tree to see if the engine would vibrate uncontrollably. It actually ran more smoothly. This was my first successful flight, he said. (To practice eating while aloft, his dining room table and chair were suspended on cables from the high ceiling so he sat six feet up. He walked to the table on stilts. His valet handed the food up to him.)

He built his airships as he was built: small and lightweight. Le Petit Santos was 5 foot, 5 inches tall and weighed 110 pounds. He pioneered the use of lightweight materials. The balloon makers objected, saying the light Japanese silk he wanted to use wouldn't be strong enough. They tested the material; it was stronger than the heavy raffetas then in use. He wanted to build the smallest man-carrying spherical balloon ever, only in cubic meters. Balloons were usually 500 to 2000 cubic meters. They said there wouldn't be enough gas to lift him. He called the balloon "Brazil," and it flew very well. He packed most of it in a suitcase.

Santos-Dumont next built a series of dirigibles that look like flying machines right out of Jules Verne, his favorite childhood author. They were big cigar-shaped balloons, under which, in a net of ropes, hung a small wicker basket, just big enough for the little aeronaut to stand. Just behind the basket was a small gas engine with a propeller of square blades. For stability his airships had big rudders and trailed about a hundred yards of "guide rope." We know the sleek German zeppelins that came later. Santos-Dumont's airships are like the backstage of a show. Nothing is tucked in. His critics said that a gas engine would ignite the hydrogen in the balloon above. They were wrong, but he had other problems. His first flight was a success, even if it ended when the ship folded in two. "I felt nothing but elation—I had navigated the air," he said, and he quickly set to improving his design. The public named the ships after him: Santos-Dumont No. 1, 2, 3 . . . on up to 22. (Skipping No. 8, which he considered unlucky. He wouldn't even fly on the eighth day of the month.) He built a complete aerodrome with an airship hangar, workshops, and a hookup to the gasworks for refills.

He was gallant about his mishaps. He accepted the smaller accidents, regarding them "as a kind of insurance against more terrible ones," he said. On his first attempt to win the Deutsch prize, when he was hung up in M. Edmond de Rothschild's tallest chestnut tree, the royalty next door, Princess Isabel, Comtesse d'Eu, sent over a basket lunch to be hoisted up to him. On his second attempt to win the prize, he crashed into the Hotel Trocadero. His balloon exploded, leaving him dangling on a ledge. He praised the "brave firemen" of Paris for rescuing him.

In competition, he was chivalrous. To win the Deutsch prize he had to announce his flight twenty-four hours in advance, leaving him to the mercy of changing weather. He was "barred by common courtesy" from calling on the judges in the most favorable flying hours, the calm of dawn. "The duelist may call out his friends at that sacred hour, but not the airship captain!" He distributed the prize, 100,000 francs, among the Paris poor and his mechanics. With other prize money he founded his own prize to advance aeronautics.

In his smallest airship, the No. 9, nicknamed the "Baladeuse," the "stroller," he became a regular Parisian sight, a boulevardier of the sky. He promenaded above the Bois de Boulogne at the fashionable hour in the little airship, cruised the Champs Élysées stopping for drinks at cafés, his ship tied to a chair. He would stop at his apartment by the Arc de Triomphe for coffee, while two servants held his ship. He flew to luncheon with friends, visited "fellow clubmen" at St. Cloud, home of the Aero Club, and was invited to fly over a Grand Military Review. "He was the perfect ornament to the City of Light, in its greatest days," says biographer Peter Wykeham.

He was famous. His suits, hats, ties, and high white collars were copied; the collars were known as "Santos-Dumonts." Edison sent a signed picture: "To Santos-Dumont, pioneer of the air, the homage of Edison." The French press had declared his prize-winning flight a red-letter day in history. When others began building dirigibles they were called Santos-Dumonts. He flew a scarlet pennant off his airships, a line of poetry he had learned in childhood, Camoen's *Lusiad,* "the epic poet of my race": *"Por mares nunca d'antes navegados!"* "O'er seas heretofore unsailed!" He planned to build a flying house, he said, and stay up for weeks at a time. He fancied a flight to the North Pole.

There were rumors in Europe that the Wright brothers had flown. But who had witnessed it? If the Americans didn't believe the Wright Brothers, why should the French? Octave Chanute, correspondent of the Wrights, had come to Paris to lecture on some of the Wrights' developments months before their flight. No one believed the Wrights could fly, but the aeronauts began to believe an airplane was possible. Santos-Dumont and others began to design heavier-than-air ships.

Ernest Archdeacon, a prominent Paris lawyer, offered a prize for the first powered flight. If the Wrights had really flown they would compete. In 1906 Santos-Dumont won the prize in an ungainly machine. He flew for 60 meters, the world's first public flight. He flew further, 220 meters, winning another prize. "Vole, Vole, Vole!" said the headline in *La Nation,* "Fly, Fly, Fly!" The Wrights were *bluffeurs;* Santos-Dumont was the father of flight. The Wrights had made a flight of *25 miles* the year before, but no one believed it. Brazil placed Santos-Dumont's face on their money. Paris set his name into memorials. He built another airplane, nicknamed the "Demoiselle," the dragonfly, for its transparent wings of Japanese silk. It was small enough to fit in the back of a touring automobile. People wanted to buy Demoiselles. He offered his plans for free, just as he had done with his airships.

Other Europeans built airplanes; then Wilbur Wright arrived in 1908. After months of delay rebuilding a damaged machine, Wilbur flew in four sweeping curves and the French were awestruck. He flew figure-eights. We have not begun to fly, they said.

Once Wilbur Wright flew, Alberto Santos-Dumont was prehistory. The gallantry of his endeavor with poetry on banners, royalty sending up lunch, fashionable promenades in his little balloon, and his talk of honoring Brazil, belonged to the rapidly fading Belle Époque.

He gave it all up in an afternoon, his Paris aerodrome, and one he had built in Monaco. He gave up his balloons and his airplanes. The word around Paris was that he had suffered a nervous breakdown. His first English-language biographer says that Santos-Dumont was diagnosed with multiple sclerosis. He never flew again.

Le Petit Santos was soon forgotten in Paris. He would revisit Paris, but in time even aviators didn't recognize his name. Brazil hung on to its hero of the "XX Century," arguing into the 1940's that the Wrights had no claim on the first flight.

Depressed, given at times to weeping melancholy, he restlessly moved between Paris and Brazil. Or as he had said of being lost in fog in his balloon, he entered a twenty-two-year-long "limbo of nothingness." He was, as he once said, a prisoner of space.

At the outset of World War I, living in a French seaside town, he was accused of spying. He was on the roof day and night with a telescope. The local police searched his house. Santos-Dumont was outraged. He burned all his papers, all the scientific data, diaries, and letters, and returned to Brazil.

News reached him that zeppelins were being used to bomb cities. He despaired. There was "real mental conflict between the thing created and its creator," says a Brazilian account of this "genius victim of his own talent." He had written in 1905 that the "first practical use of air-ship will be found in war" and that airships would menace fleets, destroy submarines, and rout armies. Now that it was happening, it caused him unending anguish. He held himself responsible.

He called aviation "the daughter of our watchful care." His descriptions of balloons were loving, perhaps erotic. "We set foot on solid ground, and stood there, watching the balloon die, stretched out in the field, like a great bird that dies beating its wings," he wrote of his first flight.

His sorrow had more depth than we, at this distance, can measure. "He now believes, that he is more infamous than the devil. A feeling of repentance invades him and leaves him in a flood of tears," Martin du Gard said. His friends tried to hide the news of any airplane crash from him.

He bought a house, called the "Encantada," in the mountains of a summer resort forty miles from Rio, and there kept to himself, taking short walks, observing the night sky, tinkering with gadgets. He was, say his countrymen, "the sad man of the Encantada," the enchanted house.

In 1926 he wrote to the League of Nations, asking them to ban airplanes in war. The League was discussing limiting other armaments. "It is known, nevertheless, what flying machines are capable of doing: what they did during the last war, enables us to foresee, horrified, the degree of destructive power to which they may attain in the future, as sowers of death, not only among the opposing forces in the field, but also, unfortunately, among the inoffensive people in the zone behind the battle line."

He offered "a prize of 10,000 francs for the best work on the interdiction of flying machines as an arm of combat and bombing purposes."

When two aviators landed in Paris after their around-the-world flight in 1930, he wrote to them, "At the beginning of this century, we, the founders of Aeronautics, had dreamed of a peaceful, magnificent future for same, the war—what could we do?—seized on our work to serve the purpose of fratricidal hatred."

He returned to Europe, and to a Paris he did not know. His condition worsened and he entered a Swiss sanatorium. Santos-Dumont was invited to preside at the gala banquet honoring Charles Lindbergh's solo flight in 1927, but he was too ill. He wept reading the letter. The handwriting of his reply was so shaky that it was barely legible, says Wykeham.

Santos-Dumont improved and returned home. Once more, Brazil welcomed him with honors. A seaplane, named Santos-Dumont for the occasion, flew out to greet him carrying a number of the country's leading intellectuals. With the crowds cheering, and Santos-Dumont looking on, the plane tipped sideways into the water and exploded. It was a national tragedy; all on board were killed. "How many lives sacrificed for my humble self!" he said.

The tragedy haunted him. "The slightest noise upsets him," friends said. He tried to take his life. He had once found tranquillity in his beloved Brazil, but in 1930 there was a coup, followed by civil war. Brazil was the second largest importer of U.S. airplanes. The Brazilian government was buying Glenn Martin's bombers. Santos-Dumont had gone to rest at a beach house. Bombers passed overhead. He could hear the bombs falling in the distance. The government was bombing the rebels, Brazilians were killing Brazilians. Santos-Dumont hung himself. He was fifty-nine years old. Brazil was shamed into two days of peace.

"Those who, like myself, were the humble pioneers of the conquest of air, thought more about creating new means for the pacific expansion of peoples, than of giving them new arms of combat," he had said. Everything remains half-discovered. We invent the airplane, television, wonders without end, but know not what use to put them to.

In his last days, Santos-Dumont had been looking for the secrets of the birds and maybe their grace as well. He had risen from his depression for a time and began to experiment again. In his youth he had said that no one would get into the air by flapping like a bird. He knew that from studying the plantation machinery. But now as "he studied deeply all that has to do with birds" he had a change of heart. He built wings from swan feathers, wings connected by wires to a central motor. With these wings, he said, a man can fly free like an eagle, fly over the highest mountain. It was as if he were trying to find his way back to that childhood game. Who flies? Does man fly?

"Exercise common sense," a friend had advised him in his youth "Does man fly? No. Does the bird fly? Yes. Then, if man would fly, let him imitate the bird. Nature has made the bird. Nature never goes wrong."

II

Flying is our newest antique. At the start of the century, no one had flown in a powered airplane. Now airplanes have changed our lives, but not in the fantastic ways once promised. We have not become birds—bird-men as they said at first. We have become passengers in search of luggage, pilots circling airports, pilots bombing cities.

The miracle of bird flight is so much greater than this: flying by sound, by memory, by stars, by magnetic pull, from a backyard in North America to a tree in a South American rainforest.

Near the end of his life, Santos-Dumont had returned to the original dream of flying like a bird, the dream that awakens when a Wright Brothers wing moves and flexes. Recalling those Wright Brothers photos of a glider skimming the dunes, I signed up for hang gliding school. Hang gliding is early flight. Before the Wrights, Otto Lilienthal in Germany was world famous for his hang gliding with wings shaped like "the out-spread pinions of a soaring bird." He was killed as a result of a crash in 1896. The Wrights pursued a different design with their gliders.

To run downhill with a large wing on your back is to join a large fraternity of the brave and the reckless, of visionaries and fools. You are in a brotherhood with test pilots and backyard crackpots.

On a humid day a dozen of us are running up and down a steep hill trying desperately to fly. We are strapped into harnesses and helmets. When we first pick up the wing, and point it into the wind, it starts to dance, promising easy flight. But the glider is flirting with us. By day's end we will have more in common with moles than eagles.

The teacher, Dave Baxter, runs along with a bullhorn yelling: Flare! Don't let the kite get ahead of the pilot! Lean forward! Lean! *Speed! Speed!* And then he spins around to those awaiting their turn: Did you see that? What did he do wrong?

Baxter urgently wants us to fly in his "kites." He is tough and almost evangelical: He wants to baptize us in the air. He has flown for twenty years. On his longest soaring flight he was aloft for four and a half hours, reaching an altitude of 10,200 feet.

He has an assistant who sometimes runs alongside trying to steady your wing. There is a strong encounter with your youngest self. Someone is running alongside as you try to maintain balance. The last time this ever happened my father was teaching me to ride a bicycle. I can remember the exact spot where I looked behind to see if he was still holding me level; he was standing far away. I was on my own.

We only have to learn five basic steps, Baxter tells us. "Then for the next two or three hours you can't believe the bliss you're in." First, balance the wing as you pick it up at a twenty-degree angle of attack. Second, pick a target. Look ahead; avoid the tendency to look down or up. Where you look is where you go. Envision a runway. Third, take three steps: walk, jog, run. Avoid pushing the nose up. Now "the magic starts," he says. When you reach seven to eight miles per hour the glider is off your shoulder. Fourth, tow the glider with your harness. Lean in, run and pull with the harness. At eight to fourteen miles per hour you'll be weight-less—"the fastest weight loss program we know." And fifth: Keep running. You'll be able to take incredibly long strides. Moon walking. "Fifty percent of people lose their mind they're so yahooed!" he says. He adds a sixth step: Relax!

"When you do it right—it feels right—it feels like you've been there before. You feel very connected to the equipment." Sounds simple.

With the eagerness of the born-yesterday, we proceed to write an encyclopedia of how to do the six basic steps wrong. We look like those early silent film clips of the earnestly crazy flapping their wings trying to fly, or bouncing along in a thirteen-decker plane that collapses. We are pulled along by the kites like rodeo cowboys thrown from their horses. We are dragged downhill on our knees look-ing prayerful. Our class mows the grass low. Baxter calls it "grass surfing." At day's end he lines us up to judge the "grass surfing" champion. One woman has grass stains on her shoulders. We have driven his gliders into the ground like giant lawn darts. We have broken two gliders, bending the leading edge of one and the down bar of another. They are hauled off to the shop for quick repair, just as in the days of early aviation.

The Bleriot flight school in Pau, France, in 1912 may have shared this spirit. "The series of attempts, expectations, disappointments, extends over seven days, and seven dull days they are, during which uncertainty is added to physical exer-tion," recalled Jean Conneau, a French naval officer. "From time to time an inci-dent breaks the monotony. . . . Over yonder there is a machine rolling in a fantastic manner. Here a pupil alights, his face is black, his clothes saturated with oil, give him the appearance of a demon; he takes them off, but leaves behind him as he goes an unbearable smell. I assure you, learning aviation lacks poetry. Some-times I would closely inspect the aeroplanes which are put at our disposal. Each of the parts [has] been more than once replaced. One has a brand-new right wing and a dirty, oil-stained left one. Another machine is all the colors of the rainbow."

I am having trouble just balancing the fifty-pound glider on my shoulders—it keeps slipping off. The glider and I have wrestled several rounds. The glider is try-ing to get me to fly, but I won't listen. Calling it a kite, a children's toy, only adds to the frustration.

As you run along it seems possible—you can feel the glider talk to you—every-thing seems to be happening faster and higher than it really is. There is a lot that seems to be happening—or about to happen. It's like a car skidding in all direc-tions, is how the Englishman Charles Stewart Rolls described his first time aloft in a Wright machine. (He was later killed in a crash.)

I'm running and as the glider lifts me up, I feel as if I'm being pulled up by the scruff of my neck and my feet are windmilling like a cartoon character. At that moment I think: Oh-my-God-this-will-take-off! I could fly—I'm flyin'—I come thunking down. The ride is over. The instinct is to stop running, not to leave the earth. After all that effort, one wants to stay, if only to see how things turn out.

I have recreated another aspect of early aviation: the disbelieving crowd. "There wasn't anybody there who believed an airplane would really fly," Beckwith Havens said of his first air shows. "But when you flew, oh my, they would carry you off the field." After a day of this I will need to be carried off the field.

I have to get used to this idea: This will fly—I will fly—that is the surprise of hang gliding. You can run down hill and step off into the sky. The only thing holding you back is yourself—a burden of truth anyone would want to leave on the ground.

Sitting down to catch my breath, I review my mistakes. I wasn't calmly balancing the kite. I didn't run through the harness. When the kite started to lift I had no idea of how to control it to keep the nose at the right angle. I was a passenger, not a pilot.

By mid-afternoon, we have burned off the morning's eagerness. Discouraged and fatigued, most of the class now sits down between attempts. We look as if we have been used for tackling practice by a football team. Even our instructor looks discouraged. He probably thinks that we are a bunch of knucklehead grass surfers who belong on the ground. But he picks up a kite and shows us once again. Effortlessly he steps off the hill and into the air. It's witchcraft.

Back in the hangar, Baxter will close the day with a pep-rally talk on the glory of flight. He is an impassioned salesman for his sport. He wants converts. He tells us how we can make it to the top of the hill—the launch for a full flight—in less than a work week, just thirty hours. We can be aloft after just 75 to 125 "flights." We have each made fourteen "flights" already, a mole's air force.

He reassures us. When you get beyond the third lesson, the flying gets more intense. It starts to happen when the launch is second nature, when we have good speed and balance, when we have poise. Less than one work week and you'll be flying, he repeats.

He tells us of the excitement of his first time soaring for hours. You can see the joy in his face. He has the whole class along for the ride. As he talks we forget the frustrations of mowing the grass, our cowboy-clown act. We believe: You take one-two-three steps and you're airborne.

He wants to close, he says, by reading from Wilbur Wright's account of gliding in 1903. As he pulls out the quote, he says: "We're the oldest form of aviation. Otto Lilienthal made 2000 flights. [He doesn't mention his death.] We are not the new guys. We're the old guys, reborn with modern materials, Dacron, Mylar, . . . aviation steel, computer- generated wing shapes. For two to three hundred years, this predates power flight. We're not just some yahoos jumping off hills."

He reads from Wilbur Wright: "By long practice the management of a flying machine should become as instinctive as the balancing movements a man unconsciously employs with every step in walking, but in the early days it is easy to make blunders. . . . While the high flights were more spectacular, the low ones were fully as valuable for training purposes. Skill comes by the constant repetition of familiar feats rather than by a few overbold attempts at feats for which the performer is yet poorly prepared."

We had come to fly like the birds. We had signed our lives away to fly. The "release from liability" form said, "I understand that hang gliding is a potentially risky activity." Sure. "I am aware that hang gliding accidents may cause (and have caused) serious injury . . . and even death." Sure. In bold-face capital letters there was something about: "I expressly and voluntarily assume all risk" . . . blah blah. Let's get flying.

Toward day's end when there is only time for another run or two, someone points out a red-tail hawk riding a thermal on the ridge. Everyone is desperate to strap back into a glider, but we all freeze. A hillside of people concentrating, as if this is the first bird we have ever seen.

We watch. We envy the wild hawk. No one says a word. We are thirsty for the secret. "I can see air," a great airplane designer had said in his youth. He had grown up on the coast watching seabirds by the hour. We want to see as the hawk sees; we want to see the lift in the air.

The hawk sails out of sight. We turn back to our wings and hoist them once more to our tired shoulders. We have only the company of machines. The birds keep their secrets.

Lenticular Clouds

Harold Humes

Harold Humes, more often known as Doc Humes, was born in 1926 and worked primarily as a novelist and educator. He cofounded the Paris Review *literary magazine with Peter Matthiessen and wrote two acclaimed novels:* Underground City *(1958) and* Men Die *(1960). In the mid-1960s, he stopped writing and became paranoid and sometimes delusional.*

The following excerpt is taken from an interview conducted in his later years. In this piece Humes talks about lenticular clouds, cloud formations that resemble UFOs and that can remain stationary, lingering in one place for hours. Humes sees a relationship between these clouds and the dawning of a new age.

Toward the end of his life, Humes took on the role of a self-appointed educator of youth, mostly at and around Harvard. He passed away in 1992.

The story in a nutshell is this: when lenticular clouds were first sighted in the fifties, the problem nobody could get around was the fact that they held, that they were able to hold their position against prevailing winds. Now a normal cloud moves with the wind because it's part of the wind. In other words a cloud is water vapor, coming out of a solution it's condensing to make a cloud. Water vapor is one of the gases that makes air. When water vapor comes out of the air to make a cloud, the cloud moves along normally. It's perfectly logical. Common sense that tells you that.

The puzzling thing about lenticular clouds, the thing that puzzled everyone who looked at them, from airline pilots to meteorologists, was the fact that they could hold their position against prevailing winds. For example, you'd see a lenticular cloud sitting over the Charles River and there may be a prevailing breeze of 25 knots blowing but the cloud would stay in exactly the same position. The wind goes on, but the cloud stays put. That presented a lot of problems, because if the cloud's not moving with the wind, it suggests that it's not an ordinary meteorological phenomenon. There's something going on that's new. Definitely novel. It suggests that the cloud is not a meteorological phenomenon but in fact an ethereal phenomenon.

The idea that space is empty is something that everyone has accepted for the last one hundred years or so, but there's no reason why that has to be. In the seventieth and eightieth centuries, it was discovered and proved that light was a wave propagation. So the question arose, so what is light making waves in? For example, you can set up a glass bell jar and exhaust all the air out of it and a beam of light will still go through it, so obviously it's not making waves in the air. If light is a wave propagation and it can go through a vacuum, what in the world is it making waves in? So came the idea of the luminiferous ether, as it was called— luminiferous, meaning "light bearing," right? And ether—not like the medical ether—it doesn't have anything to do with the ether you use for anesthesia but an earlier use of the word, meaning a very insubstantial substance, an essential fluid, something very thin.

For about two centuries the ether was more or less an accepted idea of physics. There was no such thing as empty space. It was filled with the luminiferous ether. And that sounds a lot like the old theological idea of the spirit. The Holy Spirit is supposed to be a fluid that fills all space.

People have always puzzled over where you go when you're dreaming. You're obviously conscious in a very real sense, but you're obviously not in the body. The idea of an ethereal place of existence was a very attractive idea to the religionists, and for a while science and religion got along pretty well with this idea of the ether and the spirit being more or less the same thing. It's a little bit like being a fish. For example, if you were a fish you'd know a whole lot about bubbles. They're round and they're shiny and when you let go of a bubble, it falls, and for a fish it falls up.

It's possible that the atoms and molecules that make up physical matter are like bubbles that hold together and make up things. The body is made of a whole bunch of molecules that are like little bubbles, but the real thing may be the spirit of the fluid that surrounds us. What are these little bubbles in? Well, a fish would be the last to discover the existence of water. A fish would know a whole lot about

Lenticular clouds. Courtesy of the NOAA Collection

bubbles, but if you're living in the stuff you wouldn't know anything about it—unless you were a flying fish, of course, and got out of the pool every now and then, but basically speaking a fish wouldn't know anything about water.

The thing is, there probably is no such thing as empty space. Space is fluid. All of what we think of as empty space is not empty at all. It's filled with a fluid, and the particles that make up ordinary matter are tiny bubbles and droplets in this fluid.

That's where the lenticular clouds come in. Up to the time lenticular clouds appeared, about ten or twenty years ago, there was no way to perceive the spirit. You couldn't see it. It's colorless, odorless, tasteless. It doesn't have any physical appearance, you see what I'm saying? But a cloud that can hold its position against prevailing winds. . . .

Well this is probably why the Indians called them spirit clouds. Rastas call them Ja clouds. It's one of the signs of a new age. If you look in the Book of Revelations it speaks about shining clouds as being one of the hallmarks of the new age. They shine, and at night they have a pale glow to them. They're really beautiful.

That's the whole point of this thing, that these clouds are more then just a novel phenomenon. I've been watching them for twenty years, and I can tell you they're definitely novel. You can see them more and more often. Ten years ago, you might see one or two sightings a month. But their numbers have increased dramatically so that hardly a day goes by that you don't see a few in the sky. Especially the ones that are shaped like flying saucers. If you see them low on the horizon, they look exactly like a flying saucer or a ladies' hat. You can see why people call them flying saucers.

The media, the politicians, governments are always reluctant to admit the existence of anything they can't slap a tax on or stick in a box. Something that can't be immediately explained tends to undermine their authority, 'cause they're supposed to be able to explain everything. So along comes something they don't have the answers on, and they tend to dummy up and put a bag over their head, pretend it's not there. You know, when Pope Urban VIII refused to look through the telescope of Galileo, it was because he didn't want to admit the fact that Jupiter had moons on it. His argument was that if he, the Pope, who was the top dog at the time, the top tweet in the birdhouse, if he didn't see it, it didn't exist. You've got to remember that the word *real* originally meant "royal." It still does in Spanish. In French, *montreal,* Montreal, is "mount royal." If the king didn't see it, it didn't exist. If the court chronicle didn't record it, it didn't exist. When John Foster Douglas refused to recognize mainland China, he was pretending that the quarter of the world's people living there didn't exist! It's like an ostrich sticking his head in the sand, you see what I'm saying?

This nonrecognition syndrome is a symptom of anxiety and neurosis. Anxiety neurotics don't like to see anything that upsets them. So they pretend not to see it.

The cloud phenomenon is interesting because it's there for everyone to see, and kids see them all the time. The kids have no blocks on this stuff; they're not conditioned like the other people are. So the kids go out and see shining clouds doing all these incredible numbers, playing cha-cha-cha in the sky. They spot them right off the bat and they have no problem with them. It's the older people who have difficulty recognizing that we're actually moving into a new age. This is being seen by people all over the world. There are a thousand things that tend to show this.

Repressive measures are the last gasp of the pyramidal hierarchy, measures by which the whole materialistic, mechanistic culture tries to keep control. That kind of control mania is a symptom of anxiety. And it's not going to succeed. They can collect all the transmitters and shut down the phones, but people can still communicate. It's global communication that makes the difference, I think. People can get the word around, whereas thirty, forty years ago, they couldn't.

The spiritual revolution of the sixties is a real revolution. It opened up a whole lot of people's minds—everything from people rediscovering marijuana as medicine, to rediscovering music. This is a banner thing, the first time in the history of the planet that this has happened. One of the other signs mentioned in the Book of Revelations is a new music—shining clouds and a new music. Do you realize that rock and roll went all over the planet in less than twenty, thirty years? And jazz too, improvisatory music in general. Improvisatory music is different from the old-style music, which you first had to write down and then you had to rehearse. Then you had to get a whole bunch of dudes dressed in the same uniforms and a guy with a stick out in front. A big symphony orchestra is like a military organization. Modern music is created and performed in the same instant, which makes it generically different from the old style of music. So one of the predictions in the Book of Revelations is that there would be shining clouds and new music. So here you got the shining clouds and the new music.

I'll tell you an interesting thing. There's a place in Rome called the Spanish Steps, where all the kids gather. You've got kids from Poland, Africa, France, Germany, and the thing that is amazing is that they all wear the same clothes and they all listen to the same music. There's less of a barrier between two kids—one from Greece and the other from Algeria—than there is between their own parents at home. The generation gap is a bigger deal than the language barrier. Now this is the first time in history that that's been true. We've entered a modern age, and as Mao Tse-tung put it, the masters of the modern age are the children.

That's what the lenticular clouds signify. It's like a sign. It's like what they, in the old days, in biblical times called a sign—in the days when there were no books, you had to teach things one on one. Mohammed had to teach a few people, then he taught a few more people. But even though there was no Eyewitness News, the word got around, and everybody knows about Mohammed. If it's true, the word gets around.

The point is to get ideas across to people without getting hung up behind an ego trip. A lot of people run around thinking they're the prophet Isaiah or Jesus Christ or whoever. Well, yeah, you're the prophet Isaiah, you are Jesus Christ—everybody's a star, that's the whole point. Mao and Martin Luther King were sort of like early blooms in the garden; they saw it before anybody else.

In the Book of Hebrews, it says that Jesus speaks of himself as being the first of many brethren. In other words, the idea is that everybody is supposed to come up to his level someday. It's the idea of brotherly love. When the idea is extended far enough and the whole world gets into it, then everybody arrives at the level of enlightenment as the original teachers. In the Latin, when it speaks of Christ returning to earth, it uses the term *cum,* meaning "with," comes with clouds. Not on clouds. In English, it's often translated as coming on clouds, which sort of makes it hard to believe. Jesus, you see, is not a one-man show. It's a "here comes everybody" thing, you dig what I'm saying?

Marsha Cottrell, *UNT.WT.5.7* (detail)

Of Aerial Plankton and Aeolian Zones

David Lukas

The child stops, eyes ablaze in wonder at the floating umbrella drifting on the breeze. From a halo of filaments that sparkle in the sunlight hangs a tiny dark pod. A universe drifts by. The child's hand reaches out, pushing the air and the parasol of down, angling it across the field. The child watches it lift over a hedgerow as if consciously evading touch. Sometime later, perhaps hundreds of miles from the child's field, the floating dandelion seed finally descends and begins its new life.

A moment of nearly touching one of the planet's great biological mysteries occurs in the moment when a child reaches out his hand. Visible here as a seed in transit but extending into a microscopic sphere beyond comprehension, this mystery permeates all air—even in our offices, factories, and homes.

Take a moment on a sunny day to investigate the air around you. A sunbeam angling into the room is filled with motes of dust. The backlit abyss of a canyon shimmers with gauzy fibers of floating spider silk. Look carefully into the sun's halo as it slides behind a tree trunk, and you'll see the air swirl with glowing insects.

Air is far from empty. In fact, our planet's atmosphere is a dense stew that scientists are only beginning to sample and understand. This organic brew contains viruses and bacteria and all manner of plant and animal life-forms in an astonishing variety—caterpillars, spiders, aphids, butterflies, moths, beetles, mites, and other invertebrates; plus countless seeds, spores, and pollen grains of fungi, algae, mosses, liverworts, and flowering plants. The smallest members of this floating sea are microscopic, passively floating viruses and bacteria that have been given the collective name *aerial plankton,* and their numbers are incomprehensibly vast. One estimate figures there are 50 to 100 million microorganisms in an imaginary cylinder the width of a small dinner plate. Another estimate figures there are over 66,000 in every square inch of air.

Only slightly larger are the airborne spores and pollen grains of the plant kingdom. On windy days, it is sometimes possible to witness pollen so dense it forms yellow clouds or to find pollen grains coating flat surfaces or mud puddles. Spores are not so prominent in people's minds but are equally prolific. A single four-inch mushroom discharges 16 billion spores at a rate of 100 million per hour over several days, and a single puffball produces on the order of 7 trillion.

Larger, visible insects don't have a collective proper name, but they are no less important. One survey made by towing a net from a biplane came up with an estimate of 25 million insects over a square mile of the earth's surface. At times, aphids alone may reach densities of 15 billion in the same space. Single concentrations of locusts have been estimated at 10 billion individuals, and a migrating swarm of 3 billion butterflies once passed an observer in California.

What's going on in the air? Early observers and a contingent of skeptics—even to this day—have wondered if the smaller aerial plankton simply represented the passive or accidental transport of lightweight objects. At best, they said, these aerial objects were early life stages of plants and animals merely dispersing to new sites and only temporarily aloft. Evidence now suggests that both intentional and unintentional participants comprise a unique aerial ecosystem of complex interrelationships. Together they rise and fall on atmospheric tides, circling the globe, sometimes to great heights. No one knows for sure how high life travels in the atmosphere, but in 1974, Russian rockets collected air samples that contained common microbes at heights of thirty-five to fifty miles. And no one knows how long individual organisms stay aloft, though it is known that even tiny aphids can descend at will, against adverse winds, to reach favored food patches.

Although the majority of aerial travelers are either too small or too distant to be observed by humans, we occasionally glimpse the dazzling fecundity overhead. Curious to examine this in more detail, I recently scanned the sky with binoculars while the afternoon sun hovered behind a tree. To my amazement, the sky was ablaze with darting luminescent jewels of backlit insects, layer upon layer, weaving together as far as my optics could focus. They seemed to stretch to the stratosphere, but of course, I was only seeing those in the closest hundred feet or so.

There was no order in this image. The insects ranged from tiny specks of light to large bumbling beetles, some riding the wind currents like sailboats, other darting back and forth. Because they were all backlit, it was like watching a fantastic light show of zigzags, curve balls, and quick dashes. Among it wove a ceaseless tapestry of billowing spider silk.

Ballooning spiders are one conspicuous component of this airborne community. These are spiders, typically young ones, dispersing in search of new homes. Each spider rides its own silken balloon (technically, a parachute) constructed by reeling out filaments until the wind fills it. This is an effective way of traveling: airborne spiders have been reported at heights of seven miles, and they frequently land on ships hundreds of miles at sea. Occasionally the air becomes so dense with ballooning spiders that their strands completely encrust exposed surfaces and form sheets of silk known as gossamer. It's not uncommon to see swallows with spider silk wrapped on their wings, and it's a wonder they can fly through this thicket of strands at all.

Air is a complex and fickle master, at times stratified as neatly as a layered cake, other times as turbulent as a raging river. The Greek word *aeolus,* referring to the god of wind, means "varying, unsteady." For a small organism, this is particularly true because the forces of air pressure and wind velocity magnify exponentially with each incremental decrease in body size, and even small gusts represent chaotic turbulence.

In biological terms, the atmosphere has three significant layers (not counting the very high, free atmosphere, which has its own species). The aerial community uses these layers at different times for various purposes. Closest to ground, there is a band of calm air called the biological boundary layer, where insects can control their flight because their flying speed is greater than the wind speed. This is the zone where many insects carry out their daily lives. By day, when winds are strong, this band is less than ten feet in height, but in the calm air of night, it rises much higher.

Above this lies the planetary boundary layer, a region 3,000 to 6,000 feet thick where global air currents eddy and tumble across the rough surface of the planet, producing strong gusts and vertical mixing of air. During the heat of the day, great columns of hot air sometimes break out of the planetary boundary layer and surge powerfully upward into the upper atmosphere as thermals or towering thunderheads. After the sun sets and the earth's surface cools, this band shrinks to less than 1,000 feet. Despite the chaotic nature of this layer, it may serve as the primary dispersal zone for many species that surrender to the wind's fickle directives in their search for new homes.

Even higher, the geostrophic layer, free of friction with the earth's surface, is a zone of strong and constant wind sheer. Organisms that ascend at night through the narrowed planetary boundary layer can lock into this geostrophic beltway and disperse great distances with ease, but by day the geostrophic layer is difficult to reach because of its height atop turbulent lower air.

Microscopic spores and bacteria, floating seeds, and micro- and macro-insects all grapple with the ceaselessly shifting parameters of this aerial world. While tall, large-bodied humans experience wind in its horizontal dimensions, lightweight organisms encounter wind as a powerful vertical force. Even sun-warmed twigs and pebbles generate tiny updrafts, while on a landscape level, large, violent up-drafts can sweep to the top of the troposphere (five to ten miles high) in an hour. Carried aloft in an hour, a small insect may require twenty days to work its way back down to the ground, and winds could have transported it around the world to a new home by then.

Less favorable fate befalls aerial plankton that runs into fingers of earth stretch-ing high into the atmosphere. In 1802, while climbing Chimorazo, then the world's highest known peak, Alexander von Humboldt made the first written record of invertebrate fallout from the atmosphere. He noted that at extremely high altitudes, there were insects, usually found at low elevations, cast incongru-ously about the surface of snowfields, too numbed by cold ever to rise again. This fallout was the result of cold, sinking air that grounded airborne insects.

Researchers have documented this ceaseless organic rain on other mountain peaks such as in the Himalayas, Sierra Nevada, and Rocky Mountains, as well as other regions where large snowfields cool the air and cause local downdrafts that ground airborne insects. In the Himalayas, scientists counted 400 specimens dropping onto a hundred-square-foot plot in twenty minutes. Then they discov-ered something else: pedators were feasting on this steady supply of food airlifted from the plains below. There was a species of jumping spider at 22,000 feet on Mount Everest living solely on aerial plankton. There were gray-crowned rosy finches on alpine peaks in the Sierra Nevada. There were bears returning each year to a melting glacier in the Rocky Mountains to gorge on an entombed 600-year-old locust swarm. There were specialized communities of beetles and scav-enging flies living on thawing insects at the melting edge of snowfields.

After Mount Saint Helens erupted in 1980, researchers made another discovery: contrary to the popular understanding that simple plants are the first building blocks of succession, it appears that organic material and invertebrates falling from the atmosphere precede and lay down the basis for primary plant succession in a barren wasteland. Within eight weeks of the eruption, a host of invertebrate scavengers and predators (which themselves had fallen from the air) were found subsisting entirely on this fallout; none survived long, and their bodies became nutrients for the first plants. By the second summer, forty-three species of spiders alone had been found in the blast zone, all arriving by air from at least twenty miles away.

In the 1960s, biologist Lawrence Swan named the alpine region where organic matter arrives solely from atmospheric fallout the *aeolian zone*. This is a zone where animal communities carve out a living on icy snowfields or rock outcrops and traverse the snow's surface in search of fresh refrigerated food.

This evocative term could as well serve, by metaphoric extension, for the aerial zone where airborne organisms travel and find their home, a zone for which no name currently exists. Without a name, how can we speak of this important phenomenon or of the place where aerial plankton are located?

Aerial plankton and aeolian zones are now subjects of serious scientific inquiry. Entire web sites and international conferences focus on subdisciplines such as aerobiology and radar entomology, and for good reason: allergens, insect pests, and epidemic diseases are also members of the aeolian zone. In 1918, the great influenza pandemic that left more than 30 million people dead followed an erratic pathway around the globe best ascribed to wind patterns. Lyall Watson, in his book *Heaven's Breath,* suggests that all plagues and epidemics are atmospheric in origin. Earth is not "a sealed spacecraft, isolated from the environment in a convenient bubble of air." he writes. "We travel rapidly through space and time with our windows open, constantly exposed to the complex ecology of the galaxy and all it contains."

As I sit on my porch and write, I sense that the air around me is full of life. It's a sunny afternoon, and a light breeze drifts across the Central Valley of California and gently ascends the west slope of the Sierra Nevada. In the Andaman Islands, there is a wind that natives call *biliku,* "the spider," for it wraps itself about everything. I catch myself watching small insects dart through a beam of sunlight while an animated fly buzzes by my head. I take a deep breath, wrapping the air around me, filling my lungs with spores, pollen, and microorganisms—the stuff of air.

The Bering Bridge

Roald Hoffmann

The old men say
the sky was once so close
that if you shot an arrow up
it would bounce back at you. The sky
swallowed birds. Sometimes it lay
like the luxuriating fog
just above our tents
and a man could climb
to the opening at the top, where the smoke went out
and talk to the gods.
Then the redwoods came, sacrificing
all to the main trunk, and
they jacked up the sky,
and then men with balloons and telescopes
pushed it back further,
so it became difficult to talk straight to the gods,
one had to yell, or use the intercession of shamans.
Now I have flown myself across the Pacific,
seen the deep sky blue at 30,000 ft.
They say a man has walked on the moon. They
say the earth is getting warmer.
I see smog, the sky coming back down over California.

Aerial Imagination: The Character of Air in the Spaces of Le Corbusier and Alvar Aalto

Sarah Menin

> *Is not air the whole of our habitation as mortals? Is there a dwelling more vast, more spacious, or even more generally peaceful than that of air? Can man live elsewhere than in air? Neither in earth, nor in fire, nor in water is any habitation possible for him. No other element can for him take the place of place.*
> —Luce Irigaray, *The Forgetting of Air*

Prologue

Two aging, enfeebled, yet defiant egos of twentieth-century architecture are sitting in their retreats in the gloaming of their life—one in the rabid heat of the Mediterranean, one in the biting late winter of central Finland—recalling how they conceive and create the atmosphere of their built places. Both are breathing with a little difficulty. It is their heart, one way or another. Le Corbusier, pensive, looks up from his painting to squint through the golden section of a precisely framed window vent, out across the blue stretches of the Mediterranean. The air without is so hot. Within, it is mediated, controlled, and pleasant.

His sometime friend Alvar Aalto turns, in a gesture of convivial flattery, and glances across the partially overgrown, partial ruin of a courtyard, through the trees and out across a cold lake in central Finland. The air without is so cold. Within it is mediated, controlled, pleasant.

The gaze of the old men goes in and out of focus as memory and creativity dialogue. They think about each other.

Act One
Le Corbusier, Hot Air, and the Search for Unité

Setting: Hot, clifftop view over the Mediterranean, South of France

After experiencing the essence of the monastic life at Ema, the taut young body of Charles-Édouard Jeanneret adjusted his appellation to take up his call. But he is momentarily undone, gasping asthmatically. Uncharacteristically lost for a response—a physical crisis is called up. Then—upright, so very upright—he once again makes peace with breath and takes to the air. Flying the nest, Jeanneret assumes the persona of Le Corbusier, making sweeping statements in public with ease. Becoming known as Corbu to his friends, he flies (he thinks) away from his roots. And into himself.

Memories nonetheless hang in the spaces he creates.

Growing up, the atmosphere was as cold as the preaching of his Mother was fervent. The practice of the Father was as clockwork as the favor for the brother was palpable. The youngest, Corbu's passion was as Other, as his will was defiant. Chased by consequent aloneness and isolation, he drew life higher and higher into his head—fantasy began to pursue him, calling him in from the mountains and the trees. The boy bent over in pursuance of Truth, seeking to reside there, in lieu of being reunited with something Other—still preaching.

Perched now, aged on a cliffside, hot, arid, and dusty, is a wooden cabin, a temporary cell in honor of the monklike Père Corbu. This Petit Cabanon, built by that great and aged ego as a retreat from the everyday, is becoming his retreat from his struggle of a life in Paris. Crickets rub knees with the aged father of ineffable space, which is the essence to which he and his spaces aspire. Outside, the unlimited space is hot. Blinding sunlight ripples over the warm water beneath.

The old man stands at the door of the cool square cell, insulated by nature, wherein the air is cool—"by virtue, of course, of the very detailed and exceptional skill of the architect"—he crones, memorially. The hot open air that is without is admitted, transformed into his space by, he boasts, meditation. This rehearsal of the idea of "dwelling," in nature's simplest garb of wood, has floored his acolytes. He imagines them asking, "But where, of master, is thy pilotti, where is thy concrete?" "Stripped," he imagines himself reply, "stripped to nature's core!" He smiles, recalling how he and his friend Aalto shared a chuckle at this—at how they had both begun to speak not of white boxes, but of something more archetypal and earthy—"stripped to nature's core," he repeats to the gentle breeze. The sky is dominant, and the view is restrained from within, squarely framed, stripped of its expanse. The timber takes the strain of my very late modernism, translating it with a vernacular accent—but one that is, ironically, more universal than those chalets of my mountain youth. He stops and glances around, as if to check the truth of his statement. Here, he continues for his own justification, "I demonstrate relationship, man and nature—a continuum. My place does not compete; rather, it completes nature. He moves over to a table, and settles down.

He settles not in his work shed but in this aphorism of a dwelling. "This space is pure," he senses. "Pure geometric form in which the spirit of Euclid sings and the body of Man moves with perfect ease." He feels himself dwell, too, in the space left by his wife's death, and with a sudden movement of the air, Père Corbu feels a constricted confusion and a tightening chest.

As he begins to gasp, he looks out the window, trying to focus on something, recalling the mountain pass he used to travel with his father who walked far ahead. He draws himself up, searching for air, which is all around but unavailable.

For a moment, he is at a loss, uncertain of the boundary of physical and mental pain. He knows how to face this confusion now as he breathes a little more easily, seeing his father go on, up ahead.

Alone. He looks at what is inside and what is outside. "Spatial games—less dramatic than not being able to breathe!" he laughs, half choking, aloud. He knows that he is good with words and images. "What is real is what I expel with my pen"—infected or enriched. It is part of himself. But it would be arrogance not to acknowledge the power of the setting, the smell, the sun, the sense of something Other. Some cloaked figure on a dusty road, far ahead on the mountain pass.

Recalling his drink-sopped ramblings during Aalto's last visit, he admits, "Why did I not tell him that I actually agreed? Truth is in buildings, not talk. I was always just playing with words. And that tongue-tied top dog can't compete on that score . . . and I can't bear to lose!" The old man scribbles furiously, inadequately. "But his sketching hand, that's another matter," he whispers.

Sketching out pictures, he thinks of home and habitations scratched in the air, his memories of pen-poured Puritanism arise, he feels an element of heated hate for the absolute to which home aspired. Control was all. "And anyway," he admits, "look at me now. The friend of Papists, if not Puritans! Not to say the Father of Purism!" His weak chest tightens at the memory. "But that defiance, both at home, and beyond. Industry was marching forward—and with it my notion of art and life and place. . . . And nature all around."

"Defiance and confinement builds space," he thinks. "It certainly formed me!" he laughs, without a wheeze—knowing it to be true only for one of his many selves. "There is always the choice. Offer an open and refreshing welcome which inspires, or cramped and dingy depression which repels. I aspire." He breathes with a little difficulty and thinks of the hill of Ronchamp, of warping enclosure, of light and aspiration, the genius loci and the four horizons. "Twisting a welcome to one's self."

But on this arid cliff, heat pounds on the skin of the Cabanon, as if on his own back. "Creative alchemy, practiced for all time as man's salvation from the reality of natural forces. That's what I can do. Sometimes the mechanics of the how are a challenge!" Corbu says a little nervously thinking of the prosecutions that followed some of the inaugurations. "Better not dwell . . ." Yet he does. "But, oh, to dwell. To dwell with contentment, Père Corbu. Dwell in nature's womb, my son. Singular and alone." With some irony, he thinks of himself racing ahead of his mother, hiding and wheezing in a bush, free, if a little breathless.

He wonders if these are related. "I have always wanted to start from zero, to be ineffable . . . " He is sure that, at least, is clear. "Some inhale my vision with intellectual and sensory delight. But like everyone, to air I owe my life's beginning, my birth and my death; on air, I nourish myself; in air, I am housed; thanks to air, I can move about, can exercise a faculty for action, can manifest myself, can see and speak," he concludes, relieved. "The a priori condition of all aprioris,"[1] adds the philosopher, continuing. "Despite what Heidegger believed, I cannot practice without aerial matter. It is with both the reality and the phenomenal character of air that I design."

"Aerial matter is the essence," he says, grinding dust into the floor with his bare foot. "Yet what is that matter? That which comes into my privacy uninvited, on the wind? Active and dynamic flights of fancy make the creative thing happen. I can't stop. (My) Life depends on my work." Pausing, he observes his feelings, waiting to hear his next words, which do not come. A feeling of emptiness remains. He closes his eyes against a rush of intense sirroco wind and the dust, still sketching, blind. His mind slips between the simultaneous dwellings of nature, sensation, and thought—momentarily editing out the heat and dust, remembering places and the mass mingling of air and ground and growth and people.

Père Corbu rises, slowly. He will descend the rocks and cleanse himself. Moving from dusty earth into clear water beneath. As he hones his trim, aged body, his mind drowns in the cool water and hot sun and warm sea air. He tries to think of shelter. But his thoughts now muddle care, caves, caresses, and he has lost the focus on his Cabanon. He moans, swallowing some salty water. He is momentarily confused—to which shore is he heading? He strokes his way toward land. "Is this

Lake Geneva? Is that Mother's place?" He is hot with aches and pains of heart and body. "Water, like air is not bound to place," he considers. "But at the same time it makes place."

He recalls Yvonne, his vestigial wife and nemesis. "Since you moved on, I have wanted isolation more." The breeze seems to answer for his wife. "Being without is a struggle that is within myself," he thinks. He struggles a little, spitting at the rock as he climbs and speculating about the coldness and emptiness and nothingness that is air—that yet ensures life—at this very moment he wheezes. And the old man is a little sad. "I play Superman—comfortable with solitude and power, but . . . " He feels a constriction in his throat as he continues his climb, taking some pleasure in the power of death.

He wants to stride back to his den of creativity but has energy only to shuffle toward his retreat, through the dust and grit, sending small lizards scurrying into the odd remnants of dry grass. His mind dwells on power, and moves (with a great deal more ease than his body) to rest on a favorite conceit, "the ineffable arrival—speechless and primal," he declares, returning to his seat. He knows that people respond to his work either with feelings of incomprehensible diabolism or they are drawn up with aerial ease, clearness, and radiancy.

He takes out a pen, attempting to find sanctuary with the action of his own hand. But these thoughts remain unspoken as Père Corbu's memory mingles with his adoration of his mother (Marie)—Jesus's Mother Mary—and the prostitute Mary Magdalene. For a moment, these three tumble into indefinable oneness in a tomb, between reality and fantasy, "between what I needed and what I got."

His mind wanders, and he seems to hear a weak son whimper, never making it into the heart that mattered. He recalls his youth, feeling passion for a bud, and the rhythmic repetition of the hills on the horizon. "I gathered into myself patterns of leaves, and the character of skeletal remains in musty museums. But it stifled me. So I hauled myself up to higher realms!"

"All my life I have invited space to promenade as between the vestiges of the Acropolis, meeting the odd god, reaching for shade, in which the air is cool. Shade was brought, thus caught, into the interior. The internal streets of my modern Acropolis are cool and dark—a positive haven from the beating of the sun god—and people may openly worship on their terraces or the sculpted roof." He feels puzzled. "Workers worshipping, did I get that wrong?" Allowing himself to recall again the fracas over the building, "This architecture cared . . . ," he hoped, " . . . with a firm hand . . . " he admitted.

From within himself, something is gripping his heart. That someone is cooling and controlling him. His squint disappears as the present fades into blurred memory. But he sees clearly a boy's squint enunciate, "I am reforming myself for you." Corbu's dry lips begin, in turn, to recite a dogmatic script: "Defying myself and others I will unearth place. In arid lands I will bring cool havens. Air will be free to comfort the body's distress as folkloric ancestors knew to do. Matter will manipulate the space to suit the man, the woman, the child, the Med or Ahmedabad."

Somehow detached, he hears his own words, but their moral urgency has slipped and fallen back into himself. Frightened of being out of step, the old man tries to recall the meaning, but thoughts streak, naked, into his stead, repeating and altering, "I am (re)forming (Myself) For you." Finding himself wandering around the images of his wife, his Mother, Mary, and Magdalene again, Père Corbu's forces himself to return his gaze to focus on reality and the wave of the olive branch outside the window. He squints toward this gesture. His chest is tight as he holds his secrets to it—his spiritual and psychological agenda thumping at his frail body.

"Health, logic, daring, harmony, perfection. Just as it should be," he admonishes himself, pushing on with his work of translating this admirable agenda from a secret one of ritual, paganism even Orphism and sexuality.[2] Whispering to the reaching branch, "Striving for the womb, I am carving out a tomblike form, yearning towards light and birth of insight and initiation . . . . Marie (Mother) I am (re)forming (Myself) for you." The branch bobs in silent, almost nonchalant recognition, and the old man worries about whether his correspondence might hold palimpsests of such passions.

The window draws him out of himself for a moment. Père Corbu articulates his achievement for no one but the wind. He feels air's flirtatious stroke around his wet neck, observing how good architecture "walks" and "moves," inside as well as outside, like air. "My boundaries are thus both creatively and enigmatically defined," he says with a smile of intrigue, which changes as he hears his own words, which will keep growing and flowing. "Marie (Mother), I am (per)forming (Myself) For (ever) (For) you." Scared of his unknown selves, the old man strives to remember his Yvonne—that grounding force of earthy comfort. This fails, so he turns to his architecture, conjuring a moment his Purist space, which like a kiss, reassures and moves him—an invitation to aerial travel. He is reassured, but he is still alone.

Père Corbu tries to invert his condition, recalling the joy of self-emptying. "Oh, with those steps into the monastery at Ema, when I was just 20, I was caught by the ascetic cells and the communal yearnings. I was infected." His disease spread to a yearning—"that cloistered contradiction between isolation and unity cultured me deeply, preying on my deep sense of being toujours seul.[3] Thought is uncovered and one must fight with it. And to find it before fighting it, you must seek it in solitude."[4] But he longs again for the union of simplicity of body, mind, and spirit of those Brothers. And Albert, his own brother—also weeps. Brother on brother's old and bony shoulder. He fiddled nervously, but never to Mother's satisfaction. He stuttered his way about her affection. But this was as good as it got, brother supposes.

He tries to think of something else, of his creative excellence, but his mind stutters M M M (Mother) I am (re)forming (Myself) For you. Together the brothers speak of Marie, and flee (Longingly).

He sweeps dust from the windowsill and continues meandering around his creations as they caress nature for him. The memorial presence of nature, and Women, leak through the time-escaping boundary, drawing his mind toward the dialogue of here and there. Woman is imprinted on his space—projected into the past to heal an emptiness. And he feels suddenly silent. A gap is filled, an emptiness replete.

Momentarily. The space plays with his imagination, and he finds himself drawn back into the play of memory—and a little boy scrambling to catch his striding Father—Old Corbu jerks back. His weak chest wheezing as the mountains loom. "Would someone just play with me?" He is no longer able to clamber after the image of his disappearing father. The idea is too much to bear. He is thankful that he has made in the physical a parallel mysticism—"My architectural promenade— a creative insurance policy!" he says without a smile.

His father touches the edges of himself where the gestation of one's imagination finishes. He recoils from the thought. From passion and pain came power and potential. Indignant, in self-defense, he cries out, "Damn it. I have brought back the temple to the Family. I restored the condition of nature to the life of man."[5] And the branch still nods indifferently. "Eyes which do not see," he spits.[6]

As the young man grew, his portfolio of potential grew, and his influence and ego grew to bursting. But now, eventually aged, his energy wanes, and he sometimes despairs—an embittered Quixote—a person apart, with wider visions and ascetic discipline. Almost comfortably, he settles into this cycle of rejection and yearning for praise—a ritual of his late years. Though Père Corbu rebukes society, he still wants to be honored by it.

With anger surging in his veins, Père Corbu feels he is being hounded, mistaking himself for others—again. His mind moves from doubt into certainty. He has manifested a kind of ascensional psychology. "My work is better for people and their minds. It defies ostentation, hungering for the ordinary. It's so intrinsically linked to aerial matter . . . drawing man to greater knowledge of himself . . ." He stops, feeling his conclusion forming, " . . . and their potential—whether they liked it or not. Aesthetics or ethics, what's the difference!"

The sun beats on his baldness. Everything becomes muffled as memories mingle with sensation. A bird song draws him back to the Jura of his youth, and the wind becomes cold, and his father is calling, "Come on, son." Suddenly the action animates, and he defies the slump of age. "Oh, what was it Quixote said . . . , 'For the maddest thing a man can do in life is to be finished off by his own melancholy.'"[7] Thus galvanized to choose his end, he consciously opens himself to the elements. Fueled by the heat of the sun, he stretches, as if limbering up. Yearning for the physical freedom of youth again, he strives out of his aged self and adopts a fragile stride into the blinding sun. Yearning for unity with Orpheus and that ancient and perennial otherness, he moves more gingerly, feet bare, toward a fatal decision. He looks to the sea, taunting fate, he muses. Will he chose this as his final ablution? He stumbles down with defiant certainty and never returns. Spirit liberated from matter.[8]

Act Two
Aalto, Cold Air, and Gestures of Amelioration

Setting: Winter in central Finland, isolated

In a very different cold atmosphere, perched on a huge granite boulder, above the still, telling mirror of a lake, Alvar Aalto is standing akimbo, posing a confident, if aged, stance. A gentle wave caresses, as if reassuring the boulder as it descends to the depths of itself. Memory ebbs and flows with every breath of a life precariously perched between the abyss of memory, the fear of death, and the vital, awesome creative ascent.

Aalto's insulation against unresolved pain is again in place as he calls to his past and hears himself. "I thought of you in my loneliness, . . . I had begged you to help me."[9] His memory is sent to a dying wife or mother, and his hands reach out to caress the image.

Aalto breathes the distinctive character of the place—its cocktail of enriched climate, ground, and people. It was such breath that filled Aalto as boy and man. Tutored by the tracings of his father's forest surveys, he acts with reverence for the context rather than in defiance of the place. Alighting from a bedridden realm of sadness, Aalto thus breathes through dry tears onto his assistant-cum-fiancé-cum tutelary goddess. Together they watch him conceive a place of refuge for himself, in which she will serve and he will welcome. The place both embraces and repels.

Aalto turns from the lake to a crowd of adoring minds, and so pronounces himself in a way that contorts his several selves into a *grand seignior* who empathizes with all, but is also all things to all men and women!

"What, in the vernacular, was a methodical accommodation to circumstance, must, these days, be consciously undertaken. The young . . . " He stops himself, surveying his invited but unwanted visitors, not wishing to seem to be the bitter man that, on occasion, he feels himself to be, and continues. "The deepest human needs must be met, right up to the boundaries of psychology and beyond." This sermon is, of course, written on his wounded heart.

Sitting, in the aged cloak of his years, he is uncomfortable. Aalto is aware that he greets the site like an old friend and says, quietly as if to himself, "Always search for changes of its ever-new nature." A friend in the company is amused by this uncharacteristic consideration and laughs, but his wife throws a concerned glance his direction.

Shifting from nature to something less heartfelt, Aalto hosts the expectations of his guests. Rising now to perform an erratic role as master, in this place of ruin made new, he enjoys the young hearts pounding around him, waiting on his every word—and recalls the studio and his power over people. "It's the nature of Superman that I took from Nietzsche," he admits to himself, "not a cold emptiness . . . ." Surveying the remnants of snow, he continues, "That I was handed by death."

He notices his wife flitting around, hosting his expectation and parenting his need, and is both delighted, and annoyed. He starts as he is asked about his unusually square plan. He is in fighting form, firing back, "Arh, but not wanting to be controlled by Euclid's fair, its form bleeds away into the floor of the forest. See this courtyard, it is born perfect, and yet encases decay. Its theme might therefore be said to be testifying to both potential and ruin. Both the man and his organic forms." His eyes are drawn through the birch and spruce, out onto the vast, smooth granite boulder on which this cloisterlike sanctum rests—and he recalls his mother's form (knowing only to brag of their closeness). The glassy eyes of age disguise the sudden thawing of a grief too great to meet. He focuses on the smooth lake beneath and sees his life reflected therein.

But guests are ignorant of this and nod, dutifully, at his wisdom. And he thinks to himself that he is talking too much about architecture, shifting unwittingly to address life, as he usually did. "I think people psychologically need security."[10] He muses aloud, moving an audience that sniffles with an air of embarrassment, as if their undergarments are showing.

A keen visitor asks whether this ruin is a cell and a home wherein creation is continued. Cells and continuing creation are compounded into a sexual metaphor by a spotty youth, who then gets the giggles. Aalto is amused too, and raises an eyebrow to join his wife's blush. "Things keep coming back to nature," he comments between large sips of wine. He has probably had too much, again, she considers. But he continues against his better judgment, which he left at the bottom of the first glass. His gaze leaves the company and begins to fall, again, on the past. His wife knows the gaze and introduces a young friend, pulling Aalto back to the present. As if on cue, he focuses, sees it is a young woman, and is wakened to some primal need to relate. Moving closer to her, he quickly begins to extrapolate the previous, devious imagery of creation, wondering out loud, for her ears only, "We should allow space to outlive its conception, ready to be renewed, to adapt, to meet the beautiful faces that enter and reflect their needs. Both man and his organic forms are ever ready to caress the need." His wife is alert, and hurt. His audience is lost, the female guest embarrassed. His wife moves away, noticed by the guest, who follows her, understanding what Aalto fails to understand.

But he is amid admirers now. He is drinking, having moved, despite causing hurt and embarrassment, from descriptions of his organic forms to loose chatter about women's underwear—alluding to a lover's, exacting the memory of his mother's. Boastfully describing their soft curves translated into a challenge to strict modern norms as smooth curvaceous form. There is complete silence. He needs transmutations of the world. He creates with vitality, and others—he believes—must need this too.

Sensing he has exposed too much, he launches a vicious aside, lashing out at faceless opponents, such as the "rootless airborne internationalism" practiced by the Frenchman Corbu, he seems to deny knowing. It is as if Aalto himself is wounded by such "inhumane" ideas as he sees them—but then he recalls the wonderfully raucous time he had with the famous friend when they last sojourned together—and not a word about architecture. But those ideas somehow cut him personally, so deeply. So he rises to defend himself with pomposity: "Mine is ever a sensitive, if modern, air." One that, when explained, somehow embarrasses the assembled young hearts, who glance at his wife.

He recalls his affection for Père Corbu. "Is this the nemesis," he wonders, "of Corbu's place?" Indeed, he smiles as he wonders if that perfect timber square of Corbu's would fit directly into his courtyard—his eyes gliding over the paved ground as it mediates between safety and the precarious natural realm. Aalto's gaze notes the lifeless cold. He sighs, then remembering his guests, says, "But life must go on, and human life may even flourish, protected." But his tone again slips drunkenly to the underskirt of his mind, more slowly and thoughtfully, "Nature beyond is unlimited, unenclosed, dangerous . . . "

Aalto strives away from the view and the entourage, to the bottle and then to the drawing board. She knows creativity is coming against compulsion, and the air will be blue, before it is made sympathetic and warm. His wife reads his mood and ushers away the company into the late snow, with profound apology and a certain fear. The light is failing, and the guests begin to gather up their expectations and disappointments, and with some relief move toward the door. Glancing at his wife, they leave, with pity, not with blame. As his wife offers courteous thanks for their visit, her young friend takes her hand and pauses, looking deep into her host's glistening eyes—but the apology finds only endless exhaustion.

Aalto has turned away from the echo of partings, tumbling back from the sight of his wife to a sensation of the lost embrace of his mother. Growing to choke the memory of his gap, he cloaks himself, again, with conviviality and an air of "Top dog." Will he kick off the comfort blanket of unspecific illness? He turns back and calls, "Come again," to the young woman, whose eyes decline.

Seeking to strut his aged stuff, unaware of his diminished stature, he thinks he is disguising his personal agenda in the architectural credo of humanizing space for the little man. "And little Aalto?" his wife wonders, seeing the young woman's embarrassment.

His wife knows the burden she assumes as Aalto revels in his renewed spirit and marriage. Indeed, experiencing the dynamic reverie of this new air, he finds a synthesis of purification and reward, both moral and physical, that has an effect upon his "lifeline." Through this is flowing a degree of psychological relief—felt by all who saw the complete catastrophe of his first wife's demise. Then has come a creative (or re-creative) reverie, a place courting nature, a new wife (a new life), which is feeding the creator as he speaks.

Raising another glass to those departing, he trembles a little. "Oh yes," he tells himself with great control. "I can be the gentlest humanist"—feeling a retching urge to create an ungraspable, perfect harmony in his life. Heraclitean, and full of flux and uncertainty. "Bridging the gap . . . ," he muses, " . . . between what I needed and what I got."

He secures praise through comparisons—judging that unlike Corbu, he feels a strong compulsion to draw nature right in to the center of being—to create an association between nature, space, and freedom. "You don't associate Corbu with freedom, do you, love?" he calls, as he scribbles the mingling of landscape and building until they are almost indistinguishable, and mumbles about those who have associated air with a quality of nothingness. He thinks of Corbu, who seemed to be searching for an emptiness in air. He looks outside. "It's the richness, not the coldness, I want to find. Pure, maybe, but something fecund—that awareness of the free moment that opens up the future." He knows, because, despite his aspirations, somewhere inside his feelings tear at the Superman.

Seated at his father's old drawing board, he continues thinking of Père Corbu. "My nemesis," he jokes to himself, remembering their last playful meeting in Paris. Aalto sketches grass steps, overgrown and nonchalant in their invitation— metaphors of a vestigial welcome, "a way of loving . . . of drawing the spirit upwards." The ways of loving he knows draw a certain tension to the fore of his being. A bridge over a loss. He begins to feel that tension again. It is an exciting place to dwell—a potential space, a transitional phenomenon merging his very (divided) bits of self. Scribbling the growth pangs of nature and the impingements of culture and vice versa, he continues to sketch with increasing fervor. He watches a contemporary aesthetic emerge—one with the strong, disguising accent of human warmth. "Is that where my child is?" he wonders—in the natural place, the people, and the furnace of himself. His mind wanders to his last visit to his town hall, when people mingled, chattering about the familiar—cold, fish, and suicide.

Memory shivers, creativity cries, a mother, a brother, a lover. Dead. And himself designing for dear life. Aalto is dazed. His wife wonders, knowingly, "What on earth enables this shattered life to facilitate the creative rebuilding of a gentler pattern of life for others?"

Aalto looks up at her for a moment and then strides, as if young, out into the cold. Deaf to concern for his chest, which he shuts away as he slams the door. He must be engulfed, taken over, mingling his pounding heart with a natural pace— his life with that of the place—and be whisked up and carried beyond. But he is cold and confused and stamps the brick-paved granite to keep warm. "Well, Zarathustra, perhaps you were right that to become great, architecture—like a tree—must strike hard roots around hard rocks."[11] The cold infiltrates his capability, and the confusion between himself and his work grows, and the place is so cold, and enveloping. He coughs, and can't catch his breath, and panics, and is again nowhere near Superman, and he accepts the gentle, younger arms of his wife that draw him back to the reality of age, and warmth.

Back within, weakened. He is angry. Pulling away from an embrace. It is so much easier to speak of buildings than of himself. She doesn't see the difference. He notices her wry smile, and wonders what is his real self. His wife is on hand, and he grasps her to him.

"In Old Corbu's work there is something secretive, something for the initiated." With this thought, he feels excluded and inadequate, disabled by his dyslexic ineffability, not seeing that in this wordlessness is a freedom to which his friend aspires. As usual in panic, a creative urgency engulfs him. With determination, he hones his technique of borrowing from nature to make a more comfortable place for man, but with respect for the time-limited nature of culture. Comforted by creativity and pragmatism, he doesn't really think Old Corbu was wrong, but he surely feels better for suggesting it.

His wife comes, and touches his shoulder. "This is all rational," he preaches, on a high. "My great play on the idea of an internal sky invites elements of outside in, but erases their hostility. The sky is the background that gives form to the whole, and its expansive offering sketches a panoramic natural backdrop within which human life may be played out."[12] The pencil flies across the sheet again. She watches as he argues with life: "Things that appear rational often suffer from a considerable lack of human quality."[13] She is silently amused but suppresses any expression. She puzzles at how, unlike his often bombastic persona, his work engenders a welcome in the gentlest possible way as she moves away. He senses she has gone, feeling again the shape of the residual gap—the roots of which stretch far back into the forest. And in some small and acutely painful way, fleetingly he comes to know that his grief and his determination to offer well-being and some sort of harmony are somehow isomorphic.

Gone. "Like others," he recalls, feeling the urge that has driven him from isolation, whipping life into a frenetic party. He is used to chasing the feelings of fear from the gaping place within into that place at his center from where he is constantly rebuilding acceptance through his work. Gulping at intoxication, he seeks to banish the external sense of remoteness—the hard winter and his inner sense of alienation. Remoteness and isolation conspire creatively—between them, composing places of protection and of mingling embraces.

Aalto feels but he doesn't know. Tightening his grip, his hand scurries across the tracing paper, and his chest begins to ache. Desperately projecting his needs onto others, he calls to his wife beyond, "Away with glib intellectualization, hey. That is not the way. A mystical element is needed. N'est-ce pas!"[14] He is fighting to find the words that match the image that is appearing—wanting to translate an unquenchable thirst for human contact into buildings. She hears the pencil now, no longer flitting, but encrusting the paper with its crumbling lead. Silently—almost hiding, with tears in her eyes, she wonders if he realizes it is his thirst.

"Oh, I will build in a warmth of form, of metaphor, of texture, . . . of intent, and of function. Always and only from the human point of view . . . ,"[15] he mutters, still more quietly, " . . . with love for little man."[16] She smiles, feeling it is all so pathetic.

He pleads, "I am so tired of running," as his pencil line bleeds. With this the all-but-manic zeal ebbs away. He is reminded, through watery eyes of age and grief, of its cost. He coughs, and then jokes, as if gulping for light relief, "Whatever calm, pure spring water flows through my spaces, is certainly facilitated by the red wine that flows through me." His wife calls back in a gentle laugh.

Aalto allows himself to see that the continual flood of erudite visitors has stopped. Only the occasional drip of interested tourists remains. He recalls guffawing his convivial way to the center of attention, and adjusting himself with every breath, and is shattered to realize, for the first time, that it is the same with his buildings—forever seeking to accommodate circumstance. Creatively warding off pain! He wipes his dribbling mouth and then cites Goethe—*Ausser sich gehen*—by way of raising the tone.

Again he feels his air of conviviality become muffled within the cloak of encroaching illness. His heart is tired of disease. There his mother, within the sarcophagus of himself, yearns to be released, and to release his wives, those tutelary goddesses who held his trembling, disabled hands. But terror grips the bed to the last. Breath seeks to release him, but he dares not relinquish it.

Epilogue

The thought arises that the two infirm men unite. Through architecture, they made paths by which we may, richly and perhaps dangerously, "walk *in ourselves*."[17] Indeed, is it not an exploration of Being?

> Is Being not the unappeared—non-apparent Gestell of air? That clearing where man lives, the Gestell of air out of which he cuts his milieu, builds his home—and where he takes place.
> —Luce Irigaray, *The Forgetting of Air*

Notes

1. After Luce Irigaray, *The Forgetting of Air in Martin Heideggar,* (Austin: University of Texas, 1999), p.12.

2. From Le Corbusier, *Vers une Architecture* (London: Architectural Press, 1982), p. 23.

3. "Toujours seul": — Le Corbusier, scribbled comment on film script, cited in Ivan Žaknić, *The Final Testament of Père Corbu,* (New Haven, Conn.: Yale University Press, 1997), p. 440.

4. Le Corbusier to his parents, January 31, 1908.

5. Le Corbusier, *Mise au Point,* in Žaknić, *Final Testament,* p. 96.

6. Le Corbusier, title of chapter in *Vers une Architecture* [*Toward a new architecture*], (London: Architectural Press, 1982), p. 81.

7. Miguel de Cevantes Saavedra, *The Adventures of Don Quixote,* trans. J. M.Cohen (New York: Penguin, 1927), reread by Le Corbusier before he died. Cited by Žaknić, *Final Testament,* p.56.

8. Le Corbusier died while swimming in the sea beneath his Cabanon on August 27, 1965. The Albigensian sect, with which Le Corbusier's ancestors were said to have been involved had a tradition of "sacred suicide," "a virtuous act whereby spirit is liberated from matter." Le Corbusier is said to have said to a friend, "How nice it would be to die swimming for the sun." Cited by Žaknić, *Final Testament,* p. 67.

9. Alvar Aalto letter to Aino Aalto, c. 1932, in Gōren Schildt, *Alvar Aalto: The Mature Years* (New York: Rizzoli, 1991), p.130.

10. Aalto to Hélène de Mandrot, summer 1941, in Schildt, *Alvar Aalto,* p. 49.

11. Nietzsche, *Thus Spoke Zarathustra,* in *The Portable Nietzsche,* ed. and trans. Walter Kaufmann (New York: Penguin, 1954), p.283.

12. After Gaston Bachelard, *Air and Dreams,* (Chicago University Press: Chicago, 1972), p. 172.

13. Aalto, "Rationalism and Man," reprinted in Gōran Schildt, *Alvar Aalto Sketches,* (Cambridge, MA: MIT Press, 1985), p. 47.

14. From Aalto's, "Speech for the Centenary Jubilee of the Jyväskyla Lycée," reprinted in Schildt, *Sketches,* p. 162.

15. After Aalto, "The Humanising of Architecture," in Aalto, *Sketches,* p. 76.

16. Aalto, "Speech for the Centenary Jubilee of the Jyväskyla Lycée," p.162.

17. Friedrich Nietzsche, *The Gay Science,* trans. Walter Kaufmann (New York: Vintage Books, 1974), p. 226.

Walter Moog, from the book *The World from Above*. Courtesy of Hanns Reich

What Is Seen from the Air

Virgil Suárez

I Glyphs

An effigy, if you will, in the form
       of a turtle, slow driven on the banks
of a broken-shards mirror-like river,
       a sip of cool water to quench
the thirst of a thousand years
       of drought sifted sand and dust.
Imagine, the turtle emerges
       to dig a hole and lay its eggs.
Aerial views reveal the vessels,
       tributaries of tree patches,
vermillion vegetation with hidden
       flora and fauna, even from this far
up the scorpion joins turtle,
       catfish, antelope, they
speak their secrets heavenward,
       the shame of unreturned answers.

II Wheat Field in Reserve—Bas Relief

At first the shapes pressed against fields
of gold wheat charm the aerial on-lookers,

those who criss-cross the heavens,
moments of meditation through thick

airplane glass. These circles kissed
against the earth-Os like open mouths,

someone has stayed up late again
to make all this wheat conform to some

earth-bound plan, the undecipherable
languages of human spirit, numb words,

the earth, though, rejoices
                              in all the back-scratch.

III Christo

Madman or shaman, astral artist, what language
is this veiled pink around these Florida islets?
        A building wrapped like a gift,
        a curtain across the expanse
of the Grand Canyon, a shutter window,
a blind to divide dark from the light,
        giant umbrellas bent on the Japanese
        horizon, their snap in monsoon winds.
The artist who claims to know the mystery
of covered surfaces, a muffled sound of a rock
        dropped into a well. Outside, the walls
        gasp for air. Next a replica of the desert,
the Sonoran or Nevada desert, dust and all,
50 feet tall origami animals, dinosauria,
        perhaps out grazing for tumble weeds,
        on the rocks and fossils of other worlds,
mirrors broken on the surface of their passing.

Will Johnson, *Mountaineering Dancers.* Courtesy of Project Bandaloop

Extinct Songbirds of Maine

Stephen Petroff

There's a poem to be done
on the bird with only one wing
—Apollinaire

I heard of them first from an old Neanderthal wanderer I knew in my youth.

On the road where the both of us lived, there was a low hill with an apple tree growing at the top. Under a moon one night, he was lured to the hilltop by a flaring of dim lights in the leaves of the tree. They were not lightning bugs, he knew. He might not have climbed the hill to see lightning bugs. There was something called for that was catching the moon's light.

"At the hilltop," he said, "where the low white lights looked like flashing eyes or flashing teeth, I saw that the branches of the apple were filled with birds. Their white throats and wing feathers caught the moonlight like silver spoons and blades, as they hopped and fluttered from limb to limb. They sang as they changed places. They were night birds, nocturnal songbirds I'd never heard before, or seen—but how to describe them? I grew dizzy as soon as I saw them.

"The moon was bright but the night was very dark all the same. Wherever shadows were laid, the shadows were all black. The birds chattered as they sang, so it wasn't easy to separate the song itself from their conversation. It seemed at moments as if they sang more than one song, but it was, perhaps, simply, that they sang a long song of many verses, interrupting it with convivialities.

"Their throats and the white feathers in their wings glimmered as they wove themselves through the branches. The white feathers seemed to hold the light, so that when they passed through the black shadows, those feathers gleamed brighter for an instant.

"As I stood still, a few steps from the tree, two of the birds came out of the branches and hovered over me a moment, and I imagined they might light upon my shoulders. That would have been a dream of mine, that they'd perch on my shoulders and sing. But they merely chattered and plucked hairs from my head before returning to the tree."

He said this had happened before I was born, and that that sort of bird is now extinct.

"What is *extinct*?" I asked.

"That means it's dead with all of its kind and never comes back," he said.

"Then it wouldn't be a bird like the Phoenix," I said.

"I think the Phoenix might be extinct," he said. "I've always until only lately had trouble with that bird. The Phoenix was a bird I could never imagine, even with the help of paintings, engravings, and a drawing from a man who said he'd seen one. Once, after I'd left a party early, a girl told me later that I should have stayed, because someone had come by and passed around a dish with the ashes of a Phoenix in it. I looked for the Phoenix in so many old stories, I ended by confusing it with the Firebird, and it became an empty hole in all my thoughts.

"Now all this anxiety is gone because I've seen the Phoenix in a dream. I too could paint a picture of it. But the best part of the dream was when I heard the bird's song, which, rendered phonetically, was this:

> O by the rarest of occurences
>   you were in love with me,
>   in a blue land
> where the light was gleaming.

"I am the only one of my sort," the man said, and it was true.

He had an old language that went with his voice, but rarely spoke it out loud. When he inadvertently used one of his old words he would grimace and laugh, his only concession to nostalgia and loneliness. He spoke all languages well, but in a way that was not quite modern. For instance, he was a man who had lived under black-gown pines, and he spoke of paths in the woods as "ways." He said "the ways" when I asked him where the birds called the net-weavers strung up their nests—said "they string them in the branches up over the ways."

(The net-weavers were songbirds that are now extinct.)

The man had no single career or occupation, but had made his living by a thousand odd jobs—this odd, for instance: he'd been the frogman going down in the water, helping the drowned people getting out of their cars on lake bottoms. He would be the person most courteous and properly solicitous in that situation.

But, to stay with the subject, he had some things in common with birds. He too "migrated" during our dark months. I'm sure, now, that his leave-taking had little to do with any need to escape the cold weather. Compared to the glaciations of his childhood, our winters really had nothing to offer him, one way or another. I have the idea that he "made circuits," and that in his travels, he was most concerned with touching upon actual places that he had touched on before.

Some of the songbirds of Maine he had met first in other lands, where they'd been wintering or passing through. The net-weaver bird, just mentioned, strung the same web-like nests wherever it went. I gathered that, to him, there was no sight more moving than the webs they wove, hanging in the fog. He said the net-weaver was a bird you would burn for, if that can be imagined. He'd heard some Christian teaching, but the part he understood the best was the one that said passions can destroy you. He agreed with the implication, that some excitement might be worth his own destruction.

My Neanderthal man had once known a Christian, a contemporary and co-religionist of Roger Williams, who was the first European to note the existence of the net-weaver songbirds of the marshes. A book this man wrote says that he had fled north to Maine from the Puritan colony in the midst of a terrible crisis of the ego. "I fled from myself," he says, for he found himself bound with "false thoughts." His "Most Grievous False Thought" ("My birth is evidence of my contemptible presumptuousness") threatened to annihilate him. His study of the songbirds of Maine saved him. "For I had been advised by some subtle messenger of my God that He would not have me preferred into His presence until I had gone looking for birds and been accepted into their presence." He continues: "I walked like someone relentlessly reborn from death to death. I visited nests and birds of the marshes most, all one long summer. What a weariness, cataloguing bird-sighs and the hues of eggshells, identifying the fibres used in nest construction."

His activity, although unscientific, was nearly yogic in the intensity of concentration he achieved. His observations, while marvelously accurate, are so minutely detailed that they are unreadable. But he is the source of certain facts that would otherwise be unavailable.

The net-weavers were beautiful birds. A few pelts remain. The wings, head, and tail were dark blue, almost indigo; the breast, a deep burnt orange with yellow sparks scattered across it. The dark wings were extraordinarily long in relation to the body, with the long ends turned up like scythe blades. Those feathers were indeed used to cut the grasses the birds gathered to make their nests. Their nests hung between trees and, like those of many another bird, were built with the skins shed by snakes. When the sun touched the scaly segments of the empty snake-skins, the nests glittered.

The song that those birds sang, phonetically rendered, was this:

> Sometimes I forget I'm
> dreaming, thinking I see
> the blent light of your eyes
> but it's a dream I'm
> living fearlessly remembering.

Nearly all the birds we're discussing lived comfortably near human habitations in Maine, close to human populations; some would not shun village life. Nearly all were considered to be as common to farmsteads, large and small, as any other wild animals that could fly. They were so good while they were here.

There was a small bird, the blue wavelet with flax on its head: when the flocks of blue wavelets flew into the fields in the morning, it looked as if lakes or horse-ponds had learned to fly. They took flight in mass, coordinated blue carpets, settling to feed on flat places like fields and housesites. They fed on breeze-borne pollen motes. With their compact blue bodies and short blue wings, they looked like licks of water—with their white, tufted heads quivering like bits of foam on crested waves, they looked good in a breeze.

Their song, phonetically rendered, was this:

> Curled like a leaf when I was young
>   every morning from my sleep
>   a little god and yellow girls
> with causeless love uncurled me.

Another bird was the hedge-monk, which, feeding on seeds, hopped like a robin, but with more flair and obvious pleasure. It was a pleasure to follow a flock of them in the late afternoon into a maple woods where the sunlight hung in loops from heavy branches and made green shadows hum, while the birds darted and swirled in play, and diving down, shot upward.

Their movements seemed styled to a rite of some kind, like the gestures magicians make with their hands. And so it seemed correctly magical, when the birds, as if on signal, took position on the tree limbs, each on a branch of its own in a spot of sunlight, and sang together their long lovesong until the shadows grew cold as the sun went down.

Their song, phonetically rendered, was this:

> O by the rarest accident
> you were in love with me:
> it was the bright seeming fact.
> You were
> smoky—ardent—impetuous—in thickets,
>
> and now I think of how once in groves
> without any music I can remember
> you were carried away.
>
> Always then the tears would run
> and you would call out to me:
> Please be at peace.

The song had a full, red sound, as might have come from a creature with a much larger and warm-blooded body. The first notes were deep sighs of pleasure, followed by a few words of general approval, leading to sustained bursts of xylophonic praise. Humid weather, summer fog in particular, amplified the liquid, emotional dimensions of the song, so that just as some sounds are muffled by mists and vapors, this music was enlarged.

One of its songs only foxes could hear. So said one of our dogs. These birds did not consider human landowners to be their hosts. In consequence, some farmers were unable to shoot, trap, or poison foxes, the birds being always near to sing warning.

There was once a bird called the field-scold, that hovered and cried all day long. Until its nocturnal activities were discovered, it was considered to be merely an obnoxious bird.

There was a blind and flightless songbird. Those people who remember having seen the seventh star in the Pleiades group also recall this bird, the blind strider.

There was a bird with two nests but there's no time to say why. The bird returned from its migration not in spring but during the January thaw, built warm nests where the morning was much like the evening, cold and wet in the woods that were swollen black and umber.

This, rendered phonetically, was its song:

No rain! No more rain!
Old woman, let me read your mind,
    as you sit clutching branches
        with black frozen claws.
    Tell me, will she see nettles
        in the lace at my neck?
    Will she shrink from the scar
        on the palm of my hand?

Now I want to say more about the blind strider, which I just mentioned. I'm speaking of a bird four to six feet tall.

The bird stood on legs that, like the mandrake root, looked like human legs, with clearly defined thighs and scaly calves and shins colored green-yellow (in youth) or green-white (in age), which resembled old bronze greaves.

Gentle in nature, it would harm only flies, which it ate. It enjoyed proximity to humans, and in its old age was often known to retire to pastures where livestock was grazed, where it could be assured of flies in plenty.

In its dotage, it sang only rarely, but when it could it did, and always in the evening, when its voice carried farther than it ever could in the hectic day. No farm ever had more than one of these retired striders, and the sound of its singing rising out of a far-away field where shadows glimmered as they stepped out of the woods in the evening, the sound of its singing when everyone else was gathered at home in the evening, was prized by all the families. Because it was so lonely, its song made the evening feel cold, but the day as a whole would seem warmer, so that the children were put to bed easily when it was singing.

These "graybeards" (and even the females were so called) and these "grandmothers" (for even the males were so called) were so common that it isn't recorded when the last young blind strider was sighted. Eventually, someone came to and announced that the birds didn't seem to be breeding, and some mild concern was voiced. There always seemed to be an old blind strider singing on a farm somewhere—one year you knew where there was one, the next you knew where one was said to be, and finally you supposed they were gone. You knew for certain that they were extinct when it was proposed by a freshman legislator that the blind strider be declared the state bird. Now you must be a greybeard or a grandmother your self, to have known anyone who had heard a blind strider sing.

This is someone's word on a bird called the diver, that lived on the lakes. After its dive, it rings as it breaks the surface, returning to the air, shooting straight into the light several feet, where it hovers, shaking water from its wings, singing as it hovers. Its cry is deep and low, as if it had brought the song up from the lake's bottom, and seems affected with an echo, vibrating as if the bird were grumbling into a barrel. The song carries over the water quite a way without any diminishment in volume.

The song, phonetically rendered, is this:

> O Comrades in arms, I adjure you,
>> unhorse you and sit by the fire.
>>> I was, only was,
>>> where I've left not a relic
>> in the heart, in the vexèd heart,
>>> of my lost beloved.

It is a companion of the loon, and their voices in duet are a very great pleasure to hear.

I imagine that, in the dimly filtering light of the deep lakes where it dives, blue, green, and black are the depths; and the bird's coloration reflects the intervals of its plunge. Its head is black, wings green, belly and tail feathers a violet blue.

There is this margin note: envious men, jealous of its soundings of the lakes, talked bitterly about the bird until it was no more.

As I say these things, I begin to feel the worst sort of emptiness: the loneliness of the wretch. I think a wretch is the worst kind of "bird" because a wretch wretchedly believes a thing to be true when others have proven that even if it were true it wouldn't matter. And he is actually able to resign himself wretchedly to it all.

When I try to tell my story to more than one person at a time, this confusion takes hold of me, and love for the lost songbirds looks like something that only a wretch would need to talk about.

Even as I said that, a small dark look slipped out of my eyes and I felt tempted to say the true things of which I have no doubt and speak of the wretchedness of this world.

Full years of "human endeavor," entire decades of "history"—which must necessarily be more substantial than songbirds—have gone, as if they were racing away from our presence. Centuries roar past, and even the latest of them is starrily, unspeakably distant. Thoughts that may not even be ours linger, as memories, like old political banners blown to rags, or clouds from old storms floating back (unlikely as that might be), our connections to "all that flies." So instead of these, I remember the songbirds smaller than the meanest spirit of any man. They never left our presence of their own will. It was a misfortune that they would not allow themselves to be driven away. The extinct songbirds of Maine, as I call them, died simply, whether it was by misadventure or cycles of human ingenuity.

I know how grotesque it would seem to suggest that their extinction was part of some larger, or supernatural, migration of spirits. But it does seem as if a chokedamp moved through us, in a certain period, and that some creatures came to their senses and fled.

It was while thinking these thoughts that I became determined to say these things. Do I in my own work merely extend the rapaciousness of my kind, appropriating creatures no one else would think to claim, and putting them all to the sword? As my friend said to me, the Indians don't mention these particular birds. I've seen creatures no one else has seen, and put them all to the sword.

The pain of their disappearance is worst at those times when we remember that most of the birds were not missed until long after they were gone. It was so: the tourist bureau was preparing a pamphlet, and half the birds had to be struck from the table of contents.

One bird built her nest in dry spaces behind small waterfalls on forested hillsides. The male would dig her a hole behind the waterfall, if necessary, then build his own nest in a tree near the stream. The female lived in the nest behind the waterfall and raised their young there. The male in the tree would sing to the female, and she would answer with an identical song from beneath the waterfall. These birds were called echoes, since it appeared that the male's song was echoing from a stream in a rocky hillside.

The birds mated for life, as some birds do, and if the female died first, the mate wouldn't sing anymore. If the male died first, the female would stay on in her nest and sing from behind the falling water. Echoing-widow springs (as they were called) were prized by walkers and wanderers like my Neanderthal man, and later they were prized by landowners. The presence of an echoing-widow cascade on one's property was given all sorts of importance by romantics and real estate agents—legends were manufactured to improve or amplify the beauty of the widow's singing. This was hardly necessary, of course, because two sounds more harmonious than splashing water and birdsong haven't been found yet.

The echoing-widow is extinct as a result of the rising market value of the land in rural areas. The males were killed by men with land to sell.

The singing was more beautiful than the sound of the human voice—argue with me if you will. Perhaps the voice of one's beloved would be more beautiful, but how can this be verified? The birds are gone, and perhaps one's beloved is too. To hear something resembling the echoing-widows in beauty, you would need to listen to your true love or to the voices of your children. When, like the widowed birds, your love and children are gone, you will need to listen to the sound of your own singing voice. Imagining that, we might think, "Maybe that would be more beautiful than anything, if everything else would only keep still."

There were birds called rain doves. Their plumage was grey, or blue or silver, on which seemed to be painted drops of rain. Just as a starling is painted with stars, these birds were painted with rain.

Because the rain doves loved sunlight, they built their nests often near the open spaces associated with the habitats of humans. Most dooryards, therefore, had a rain dove nest nearby. A clearing of any sort which took the early morning sunlight was likely to have two or more of these birds dancing in it. These dances took place in midair, and only in sunlight. The male and female took turns, one diving from above and circling the other. Their blue and silver wings sparkling in cool morning light made witnesses feel the same gladness that's normally caused by a sun-shower. The sight of a hundred or more of these birds dancing half the morning over a large field caused all who saw to insist that the world is good. Nor were the blind people prevented from making their own testimonials concerning these birds.

After the rain doves had danced to their heart's content, they would sing the rest of the morning together.

Their song, rendered phonetically, was this:

> O by the rarest good fortune
> you were in love with me.
> With love I was illustrious.

> Mice and rain made me mortal.

> How easily it was explained away,
> when you, my High Friend, smiled . . .
> that there had once been

a forest on golden hill . . .
    and they say that it all
very properly vanished.

It was the blind people especially who were exhilarated when the rain doves sang. While listening to the birds singing, children who'd been blind from birth could accurately describe the faces of dead brothers and sisters, a comfort to mothers and fathers in those days of high childhood mortality. Blind men learned anew never to be hard-hearted again after they had listened one morning to the rain doves. Blind women were said to be never false-hearted again after once hearing the rain doves sing. (And if it wasn't true I wouldn't have written it down.)

On occasion, a wounded dove or hatchling might be "tamed." It would need its freedom in the morning, for its sunlight and singing, but it could be brought into the shade of the house in the afternoon. There, very much like a cat, it would allow itself to be held and petted or stroked, while it purred. The vibration of the dove's breast was of a wilder sort than the humming of a housecat. Very ancient people (and particularly the deaf) preferred the doves to cats or lapdogs. That's all over with.

There was a bird whose song, phonetically rendered, was this:

If I had given myself to you
we would never have known death.

    This is the song of my remains.

Asked, "Where do you winter?" the bird replied vaguely, "In Oceania."

During the autumn migrations, it was attracted at night to lights that shone from kitchen windows, especially if it could sit in the branches of a lilac and look in through the glass. It might otherwise perch on the windowsill or a clothesline, any place where the kitchen light might fall upon it. On nights in late September or October, you were always looking up from whatever you were doing, to see the bird at the window, imperial as an eagle in its plumage, imperious as an owl in its bearing.

There was a bird, a small warbler, that had a silver beak. While most birds sang of love and common things, this one sang of the lands it saw in its migrations. For instance, it sang of a furiously mean-spirited thorn-tree that it knew from a certain land it visited. The tree grew so densely, and was so actively thorny, that its presence at waterholes made it impossible for any but birds to drink. Therefore this bird refused to drink. This principled abstention resulted in a terrible irritability, which was reflected in the songs that it brought home to Maine. The songs lose a great deal of their substance when rendered phonetically, so no example will be given. Well, maybe just one:

> A spider wrapped a star in a cloud
> and laid it away, to eat.
> There's no escaping a spider like that.

> I saw a ghost as soon as
> you told me that I'd come to grief

> The wind is awake.
> With tin whistles rattling,
> oak leaves're loose in the road.

This bird was a favorite of my good friend. He liked it because it was small, and because it hesitated before it sang. When the bird seemed to shrink on the branch, and you saw it move forward and bloom as its throat filled, you knew it was about to sing a large and trustworthy song. Then its irascibility and sense of honor took possession of its senses—and the song burst out.

When the last flock of the silver-beaked warblers was being shot, the hundred-eyed men with guns were awfully close to the house, and my Neanderthal was beside himself, almost hysterically serene. He'd been cautioned against interfering, but at length he went out to see what he could do. After the last applause of shotgun fire, there was quite a silence. Then the Neanderthal ran into the kitchen. His chest was speckled red diagonally. When the spots of blood were wiped away, the boss of his breast was bubbly with shotgun pellets that had slipped just under the skin.

"I've saved the last of them," he cried, carefully placing a wet bundle of feathers on the table. It was a dead bird, its head folded unnaturally down, its silver beak planted in the swelling breast feathers, giving the impression that the creature had pierced its own heart deliberately. While satisfying in a sentimental way, this last image of the warbler would be wrong. Of course, I was afraid that the bird may have still been alive when he reached for it, and that—overwrought as he was—he may have been gripping it much too tightly. But even this, if true, wouldn't be an accurate analysis of its death, would it?

There was a woman who gathered the dead flock, and she buried each bird in a grave of its own. These graves were, of course, just shallow dimples in the earth, which she laid down in a spiral pattern curving out from the foot of her apple tree. Rocks the size of cabbages were painted red and set on the tops of the graves. These were eventually scattered by men who came to mow to make hay.

More extinctions are promised, I've heard, and are even being puffed as the best way to prevent government interference in the private fortunes of thieves.

So far, the last bird to go was the black bird that wasn't entirely black. While sitting, while nesting, it was a moist-seeming velvet-black bird. The undersides of its black wings were feathered with a mixture of colors: shades of blue, yellow, red—violet on the women, a fire color for the men. When they were at play between the ground and the bough, or when in flight, they were a sight that would remind you of specific moments of great emotion. When they lit somewhere to rest and folded their wings, they again seemed entirely black and this for a moment reminded you of "all that follows."

Of the birds that you might see in this life, that one was perfectly real. Although it is gone, that bird is responsible for the belief, still common among children, that diamonds come in lumps of coal.

We know how they died. They ceased migrating—that is, one year, they stayed through the autumn, when they normally would've flown, and in the winter froze.

In late September of that last year, it was noticed, they'd begun to abandon their nests. They sat on the lower limbs of trees and were observed in the fields and on the roadsides, on the ground, pecking for seeds—but it was felt that they ate without any pleasure and without any sense of obligation. After the first cold nights, they behaved as if they'd been stunned. They tilted oddly and held their heads at unnatural angles, looking down, as they sat clutching branches with black frozen claws. They made no gesture toward flying, save extending one wing and shaking it briefly, and then refolding it poorly like a broken fan—or at any rate, as if they'd become unfamiliar with wings. Before long, when the days became cold, they were seen dropping from trees. By the middle of December, they were no more to be seen.

My friend once asked whether it might not be true that I, in the perfect likeness of my kind, took ineffable pleasure in destroying the things I imagined I'd created.

When I created a black bird in its nest, the egg shells were dark blue, like a clear night sky without stars.

The black bird's throat was dark blue with a single red spot down low toward the breast, like a clear night sky without stars and only a small red planet visible on the horizon.

The bird gave me a small dark look, like a clear night sky without stars.

My crime is Divination, misuse of the treasure of poetry.

Before making this song, I owned nothing. I was a mind lost in shadows. I was a clouded heart. In the light of the circumstances, I made my choice.

My crime is Magic, abuse of my listeners, theft of the good thoughts of my teachers.

I think of the story that follows as a counter-radiant, to be held against things that look different in the cold light of day.

The last time I was at peace (that's to say, the last time before now), I sat at the table in my grandmother's kitchen, drinking tea with her, talking to my mother's sister, a visiting aunt.

I was still in my boyhood, but I had left school, and left home, and was trying to learn how to paint. I'd been reading Leonardo's notebooks for the first time and wanted to tell my grandmother all the exciting parts. I told her the part where he says that all the while when he thought he'd been learning how to live, he'd been learning instead how to die. But what had really got my attention was Leonardo's suggestion that young artists train their imaginations by looking for pictures in the water-stains on plaster walls and other accidental blots.

As I spoke, I saw that my grandmother's eyes were dry, clear, old eyes, the eyes of a bird, and that there was no emotion or lack of emotion in them, there was no emotionalism in them. Through her eyes, I saw her listening.

My aunt's eyes were glistening. She said to my grandmother, "That's what I always admired about you, when I was a kid: you could look at a cloud or a dried-up mud-puddle or smoke coming out of a chimney, and see it like it was a picture of something. It's that imagination you always had, you could point out some smutch on the wall and make us kids see the same picture you saw. I always thought: Boy! I'd love to have the imagination Ma's got!"

She said that and sighed characteristically.

"Well, I can still do that," my grandmother said. She looked up and lifted her hand to point at the ceiling.

"Yes," she said. "You see those black smears on the ceiling. When I look at those, I see a bird flying and flapping its wings." My aunt looked up and said slowly, "Yes."

There were wayward black streaks and smudges of dust across the white ceiling, pictures of nothing, but they were transformed as my grandmother spoke, describing the flight of a bird.

"It flies straight toward the lamp, and then it gets scared," my grandmother said. "It turns right, and then it turns left, it's so nervous, it doesn't know where it should go. So it thrashes its wings!"

At first glance, the marks on the ceiling had been almost invisible and insignificant, but now they took on the character of a child's drawing—clumsily executed, as by one whose hand is numb, but passionately made. We saw the bird's tail feathers veering left, its blunt head swivelling to turn right, widespread wings with individual plumes beautifully printed, feather strokes splayed like fingers.

"I see it!" my aunt said. Her voice was low and there was a golden cunning in her eyes. "It's just like you say, it looks just like a bird!"

There was a picture on the kitchen ceiling that was, to my eyes, as much of a treasure as a Chinese painting of dragons swimming through clouds.

My own voice was hoarse and I felt suddenly slain and completely at peace. "It's all exactly the way you say!"

My aunt sighed characteristically and drew a long breath. "It's that imagination I always used to envy."

"One morning last summer," my grandmother said, "I heard something rattling inside the stove, where I'd put papers I wanted to burn. I opened the lid of the stove and a bird flew out, went right to the ceiling, bounced off and flew all around the room. It had come down the chimney and it was all covered with soot, so it left its tracks all over the ceiling."

She pointed again to the picture, a true record of a bird's flight. My uncle, she said, had happened in, caught the bird in his hands, and let it fly out the door.

"So it's not something that I made up," she said, looking into my eyes, with an infinitesimal triumphal lift of her chin. "I never had to imagine that bird. It came down the chimney."

II

Ineffable Air

Stuart Allen, *Flying Silk No. 3*, Montezuma Hills, CA, 1995

Inebriate of Air

John Olson

Air is a spiritual barometer. It cannot help but inspire. It is thin and invisible yet capable of supporting an 875,000-pound Boeing 747. It whirls newspapers across the streets of deserted towns, jangles chimes, and powers the blood. Pollen and dogwood seeds drift in its currents. It might blow across the tracks on a sultry afternoon, making California poppies tremble as it carries the scent of an unknown flower, then minutes later tumble in a whirling cone of chaos and destruction, hurling cars through the sky as if they were baseballs. Every breath we breathe is freighted with its limitless domain. Language is a creature of air. The breeze gliding over my forearm is the same material with which I speak, the substance that links people together with its ethereal pleasures and an alphabet of wind and water.

Each word is a reverie. The word *air* is a glorious subtlety. Air and its variant spellings (*eir, eyr, eyre, aier, ayre, eyir, eire, eyer, ayer, aire, ayere,* and *ayr*) all stem from the Latin *aer,* as do *aerator, aerie, aerobatics, aerobics, aeronautics, aerosol,* and *aerospace.* It is such a tiny word, yet carries such a wealth of meaning and nuance. People put on airs. There is sometimes something in the air, a mood or thought in everyone's minds, an unfixed or uncertain state. It can refer to the free space above our heads in which birds fly, or it might refer to a song as a "popular air." It is full of alluring perfumes and ghostly auras, imbuing the whole body with the gentle rhythm of breath.

As soon as the air moves over me, I feel the relief of a bright, invisible medium, a living membrane at the core of each word. In the same way a song is shaped to the spirit of its music, its diffusions bring our lungs into being and with them our thoughts and feelings. Its 14.7 pounds of pressure per square inch keep my body from exploding. It is the engine of our creation and the medium in which we live. Breathing is implicit in all aspects of our being. The ancient Greek word for spirit—*pneuma*—was also the word for breath. *Pneumatic, pneumatophore, pneumococcus, pneumogastric, pneumonectomy,* and *pneumonia* are all engendered from this initial relation.

We are never alone but laced along the air in an inspired rhythm, becoming part of us as we become part of it. Throughout the course of the day, each of us will consume between 3,000 and 5,000 liters of air, the same substance that has blown over the Himalayas and Wales and the backyards of Ohio. Air that has mingled with storms and filled the morning with blue and gold. Swung gates open and scoured the canyons of Arizona. Flapped the pages of an abandoned book, splattered rain against the windows and evoked paradise in the drifting clouds.

On February 21, 1883, a group of Eskimo men built a fire on the shore of Hudson Bay and gathered around it and chanted. An old man stepped up to the fire and in a coaxing voice invited the demon of the wind to come under the fire and warm himself. The women, meanwhile, made a hullabaloo with clubs and knives, driving the demon from their houses. When it was assumed the demon had taken refuge in the fire, a vessel of water was poured on the flames. The moment steam arose in a cloud from the smoldering fire, the men gathered around and shot the demon with their rifles and then finished him off with a heavy stone. Thus, the northwesterly winds that had kept the ice long on the coast and diminished the food supply were encouraged to go.

Efforts to influence the weather are generally as fruitless as efforts to predict it with flawless accuracy. Short-term predictions achieve success rates of around 85 percent, which isn't bad, considering the volatility of weather. Air is a gas, and the original meaning of the word *gas* is "chaos." The forecasting of any complex nonlinear system like the weather is inevitably a probabilistic process based on a set of variables and conditions, such as surface features (snow and ice cover, sea surface temperature, mountains, gorges, and soil moisture), air pressure, wind speed, and cyclical patterns. Worldwide data collection, sophisticated numerical models, and state-of-the-art computing have contributed immensely to the accuracy of weather prediction, but it is still far from being 100 percent accurate, particularly in view of today's increase in global warming.

Global warming holds a litany of threats. Over the past years, it has caused severe droughts and massive flooding, and is responsible for the gradual extinction of the world's great coral reefs. Global warming is melting the Arctic, accelerating the rise in sea level, decreasing forest productivity due to insects, fire, and disease. It is reducing snowfall, causing stronger and more violent storms, and exacerbating air pollution with higher temperatures and an increase in ozone production. With climbing heat, crop failure, and the encroachment of tropical disease, global warming has brought an increase of illness and human mortality.

Europeans were understandably furious over George W. Bush's refusal to enter the United States into the Kyoto Protocol, the international political instrument to combat climate change. Christine Whitman, administrator of the Environmental Protection Agency under the Bush administration, had indicated that the United States would be going along with the protocol. Europe and Japan took Bush's about-face as a face slap.

Bush called the Kyoto treaty "unfair" and judged its application too costly. Was this a dodge made out of favor to corporate greed—or a prudent, farsighted measure? Only time will tell. As the United States produces, 25 percent of the world's total greenhouse disappears. America bears the brunt of the burden ascribed to industrialized nations to reduce emissions of carbon dioxide and five other gases by 5.2 percent of 1990 levels by the year 2010.

Controversy over the Kyoto Protocol underscores both the vagaries of climatic change and the immensity of the problem. It also serves to emphasize how vital our relationship is to the air enveloping our world.

What we breathe is in no way a permanent feature of this planet. It is not a given. The signature of so-called fossil air discovered in terrestrial rocks and sediment in the Namib desert of Africa and Miocene volcanic ashbeds in Nebraska and South Dakota testifies that earth's atmosphere is extremely volatile. The Miocene (25 to 5 million years ago) is called "the problem climate" because of its many climatic changes. It was generally a time of warmer global climates than those in the preceding Oligocene or the following Pliocene. There were palm trees and alligators in England and Northern Europe, and tree ferns shaded Montana. At the height of the Miocene climatic optimum some 17 million years ago, deep and high-latitude water temperatures were as much as 6°C warmer than they are today, and it is generally believed that elevated carbon dioxide was responsible for the great warmth of that period. Atmospheric carbon dioxide concentration was, however, uniformly low throughout the Miocene. This evidence conflicts with greenhouse theories of climate change, particularly in reference to the formation of the East Antarctic ice sheet, when (during what was obviously a colder climate) atmospheric carbon dioxide increased. The Miocene does not match contemporary theories of global warming, demonstrating once again the mystery of air.

This is not to say global warming doesn't exist. It does. Global warming is a reality, and it is directly linked to the consumption of fossil fuels. Our inability to put together a workable theory for the causes of global warming is reason for greater worry.

Again: air is a spiritual barometer. Our inability to inhabit the world in a fashion that is sensitive toward the environment is reflected in the air we breathe. Today's air quality is abysmal. It is full of notoriously harmful pollutants such as benzene, toluene, and xylenes, which are found in gasoline; perchlorethylene, which is used by the dry cleaning industry; and methylene chloride, which is used as a solvent by a number of industries. Examples of air toxins typically associated with particulate matter include heavy metals such as cadmium, mercury, chromium, and lead compounds, and semivolatile organic compounds such as polycyclic aromatic hydrocarbons, which are generally emitted from the combustion of wastes and fossil fuels.

Aromatic hydrocarbons have to do with the formation of ground-level ozone. If you live in or near an urban center, and on a warm summer day experience greater trouble breathing, chances are this is due to high levels of ozone. This is different from the stratospheric ozone that protects us from the sun's ultraviolet radiation. Ozone is the same molecule regardless of where it is found, but its significance varies. Ozone (the name is derived from a Greek word meaning "to smell") is a highly reactive, unstable molecule formed by reacting with nitrogen oxides from burning automobile fuel and other petroleum-based products in the presence of sunlight. It is also produced during lightning storms, which is why the air has that peculiar electrical odor during a storm. This type of ozone, however, is very short lasting and does not represent a significant risk to health. The real problem stems from certain volatile organic compounds such as those produced by the shellac of furniture finishing paints, cleaning solvents used by dry cleaners and computer manufacturers, and terpenes from trees. These atmospheric chemicals linger in the air and prevent the breakup of the ozone molecule back into oxygen.

Breathing ozone is like swallowing a chicken bone: we're getting something extra that we don't really want. Each ozone molecule is composed of three atoms of oxygen, one more than the oxygen molecule, which we need to breathe and to sustain life. High concentrations of ground-level ozone may cause inflammation and irritation of the respiratory tract, particularly during heavy physical activity. The resulting symptoms may include coughing, throat irritation, and breathing difficulty. It can damage lung tissue, aggravate respiratory disease, and cause people to be more susceptible to respiratory infection. Children and senior citizens are particularly vulnerable. Inhaling ozone can affect lung function and

worsen asthma attacks. Ozone also increases the susceptibility of the lungs to in-
fections, allergies, and other air pollutants. Medical studies have shown that the
harmful side effects of ozone damage to lung tissue may continue for days after
exposure. Ozone high in the sky is a good thing. Ozone low in the sky is a bad
thing.

This breakdown of one of our atmosphere's constituents may sound overly
dry and analytical to some. Others may be more impressed by how something as
tenuous as air can contain so much complexity. It was this aspect of the atmosphere that so galvanized scientists such as Joseph Priestley and Antoine Lavoisier
during the Enlightenment. In view of this, it is important to realize that their discoveries were made possible by a new way to observe phenomena. It wasn't merely
their discoveries that are so galvanizing; it was also the fashion in which they were
made. A new consciousness, a new way of thinking and writing, were developed.
Much of this new way of thinking had to do with popularizing knowledge and
empowering the average citizen.

The scientist seeks pure objectivity but certainly isn't blind to underlying cos-
mological correspondences. "The air is full of sounds," observed Ralph Waldo
Emerson, "the sky, of tokens; the ground is all memoranda and signatures; and
every object covered over with hints, which speak to the intelligent." The same
material that makes the stars glow powers the chemicals in our brains. There are
revelations in the smell of the rain and allegories in the wind. Clouds are declara-
tions of invisible life, illustrations of powers at work, extraordinary phenomena
that scientists such as Isaac Newton, Evangelista Torricelli, and Gabriel Daniel
Fahrenheit have given the names *gravity, pressure,* and *temperature.*

Air is an ideal metaphor for divinity: invisible, it is ubiquitous; mostly benign, it
is capable of immense destruction. Manna on a summer day when the rain glitters
down in the sunlight. A membrane surrounding our planet like cytoplasm surrounds the nucleus of a cell.

Air protects, envelops, and bathes us. With each breath I take, I am reminded
that I am not a single, totally self-reliant, and autonomous entity but a biological
system intimately connected with the invisible membrane pressing in on my skin,
refreshing me, and giving me life. We do not cause air. Air causes us. Breathing is a
process that occurs without our conscious intervention. There are respiratory neurons in the medulla that are highly sensitive to the amount of hydrogen ions present in the blood supply. If the number of hydrogen ions is increased, these
respiratory neurons will "sense" it and increase respiration in order to lower the
hydrogen ion level in the blood. Conversely, if these respiratory neurons sense a

decrease of the hydrogen ion concentration in the blood, they slow respiration to move the hydrogen level to normal values.

It is essentially the air itself that causes breathing. The lungs work by setting up a pressure differential with the atmosphere. The lungs are surrounded by a water-filled tissue called pleura. The pressure in the pleura is negative when compared to the air-filled areas of the lungs. This phenomenon keeps the lungs naturally inflated. Underlying all our activity, the gentle rhythm of respiration is soothing and sustaining us. This fantastic gas enveloping our planet—its feeling, its aspect, its savor; its magnificence of winds and storms; its phantasmagoria of colors and clouds; its voluptuous breezes and boundless symphonies—is as intimate a part of our lives as our dreams. "When we have gained experience with the psychology of infinite air," observed French philosopher Gaston Bachelard, "we will better understand that in infinite air dimensions are obliterated, and that we come in contact with a non dimensional matter that gives us the impression of an absolute inner sublimation."

Air awakens flight and a feeling for movement like no other element. "I'm feeling for the Air," wrote Emily Dickinson, "A dim capacity for Wings / Demeans the Dress I wear." There is a subtle fluidity in the flight of a bird, an inexpressible swiftness, a pure delight in energy. Bachelard embodies this principle in a rapture of aerodynamic prose: "If the purity, the light, the splendor of the sky calls out to pure and winged beings, and if, by an inversion that is only possible in a regime of values, the purity of a being gives purity to the world in which it lives, one will understand immediately how the imaginary wing assumes the colors of the sky and how the sky is a world of wings."

Shakespeare's dramas make frequent reference to air. Spirits pop out of it. People melt into it. It is a "brave o'er hanging firmament," a "majestical roof fretted with golden fire," and a "foul and pestilent congregation of vapors." It is "a chartered libertine," "necessity's sharp pinch," "nipping and eager," and "rheumy and unpurged," all depending on the mood and barometric pressure of the dramatic situation. In *The Tempest*, language and air are identical. Air—as Ariel—is a being ("a touch, a feeling") that flits about "flaming amazement," and Prospero's "cloud-capped tow'rs" and "gorgeous palaces" are all made of air.

Air is synonymous with thought. The volatility of the air affiliates it with thought, with hawks and skylarks and the eternally changing clouds, with the flaming reds and sumptuous pinks of sunset or a great ribbon of gold over the sea. It is never the same, never caged, never nailed down. When we feel the wind move over our skin, it is as if we are feeling a spirit move over us, a dynamic and moving mind boiling in the clouds and trees. It is too fast for time, too infinite for space.

Air is spirit. The very word *spirit* comes the Latin word *spiros,* meaning "to breathe." To the Navajo, the wind is just about everything: a source of life and a spirit of guidance. The Supreme Sacred Wind is the Supreme Creator of the Navajo universe. As James Kale McNeley explains in *Holy Wind in Navajo Philosophy:* "The mists came together and laid on top of each other, like intercourse, and Supreme Sacred Wind was created. . . . Supreme Sacred Wind lived in light and black clouds or mists in space." A further account from *The Holy Way of the Red Ant Chant,* told by Hastiin Dijoolí (White Cone), describes how the wind informs human life: "When the winds appeared and entered life they passed through the bodies of men and creatures and made the lines on the fingers, toes and heads of human beings, and on the bodies of different animals. The Wind has given men and creatures strength ever since, for at the beginning they were shrunken and flabby until it inflated them, and the Wind was creation's first food, and put motion and change into nature giving life to everything, even to the mountains and water."

In his poem "Hymn to Intellectual Beauty," Percy Bysshe Shelley refers to "the awful shadow of some unseen Power" floating "though unseen among us,—visiting / This various world with an inconstant wing / As summer winds that creep from flower to flower,—

> Like moonbeams that behind some piny mountain shower,
>> It visits with inconstant glance
>> Each human heart and countenance;
> Like hues and harmonies of evening,—
>> Like clouds in starlight widely spread,—
>> Like memory of music fled,—
>> Like aught that for its grace may be
> Dear, and yet dearer for its mystery.

On a day when the sky is blue and the temperature nudging 80 degrees, it is easy to be transported into such raptures of breath and spirit. Shelley's words and the Navajos' Supreme Sacred Wind take on special resonance. Before I've gone a mile into my daily run, I've got my T-shirt off and balled up in my hand. The air is still, yet the motion of my body feels the air glide over my skin with a deliciously silken sensation. Sounds pass through the air like thoughts through the neurotransmitters of the brain: Bruce Springsteen's "Badlands" coming from a radio, an airplane propeller chewing its way through the sky, enough birds chirping every-where to keep an auditorium of Beethovens and Mozarts busy composing, a crew of carpenters pounding on aluminum scaffolding: clang, clang, clang. I pass along a corridor of oak trees, and my body drinks in their delicious shade. I emerge into the heat again and look to the west. The Olympic Mountains are barely visible through the diaphanous haze, and one or two sailboats are foraging for whatever breeze they can find on the unusually serene surface of the Puget Sound.

Days like today are extremely rare. Here in the Pacific Northwest, it is most likely to be cloudy. Except for a few weeks in midsummer, one becomes accus-tomed to a pattern of overcast skies. There is a sameness to it, yes, but the cloud layers differ radically from day to day. Sometimes it is a still day-long blanket of gray obscuring the sun and creating a sense of being peculiarly sandwiched be-tween earth and heaven. There are days when the clouds in great puffy clumps come scudding through like great islands and mountainous continents, and days when a high layer of cirrus fans out in long diaphanous strands and spidery fin-gers. Days when the clouds are striated and scattered or rippled with pink and or-ange. Days when the clouds bulge into a swollen ball of black and purple, flare white with lightning, and bruise the sky with thunder. Days when I've gone for a run in balmy sunlight and a half-hour later felt pellets of hail stinging my legs and arms.

Clouds are the very symbol of transformation. They change continually. One might lie on the ground a long while on a warm summer day entertained by their slow evolution from elephants to castles to Spanish galleons. They are mountains of fluff, mists moiling with the breath of heaven. It's no wonder that the most pop-ularized view of heaven in the Judeo-Christian world is that of winged angels con-tentedly playing harps on mountainous clumps of cloud. To the Hebrews fleeing Egypt during the time of Moses, the most visible sign of God's presence was as a pillar of cloud: "And all the time the Lord went before them, by day a pillar of cloud to guide them on their journey, by night a pillar of fire to give them light, so that they could travel day and night" (Exodus 13:21-22).

There are such powers in the weather—thunder, lightning, outrageous winds, and divine colors—that it is only too natural to think of the sky and air as being inhabited by gods and goddesses, spirits and demons, sprites and fairies. The sky flashes or glows or collapses on the earth in a howling whirlwind of debris and de-struction, or glides over a mountain range breaking apart in rain. The sighing of the trees is the language of angels. The sly sliver of the moon tells fortunes of rain or drought. Each delirious flash of lightning is followed a second or two later by rolls of thunder: such great grinding grumbles that it sounds as if huge pieces of furniture were being moved about on the floors of heaven. The weather ignites the imagination as surely as lightning illuminates the sky. Forms concealed in darkness flare out in stark whiteness a split second, and night is instantly converted to day. It is a maniacal revelation, a delirious spectacle of unearthly splendor. Things utterly hidden and invisible minutes before are suddenly clear and available to the mind. In Matthew, the angel that descends to tell the two Marys sitting by Jesus's tomb that Jesus had risen into heaven has a face of lightning: "The Sabbath was over, and it was about daybreak on Sunday, when Mary of Magdala and the other Mary came to look at the grave. Suddenly there was a violent earthquake; an angel of the Lord descended from heaven; he came to the stone and rolled it away, and sat himself down on it. His face shown like lightning; his garments were white as snow. At the sight of him the guards shook with fear and lay like the dead" (Matthew 28:1, 4).

Haokah, the Sioux god of thunder, would either weep with the rain or smile with the sun and is depicted with two horns on his head. He beat the tattoo of the thunder on his great drum, using the wind as a drumstick. In keeping with a supernatural power, the Sioux imagination reversed the natural order of things: heat affected Haokah as cold, and cold affected him as heat.

Chinese deities connected with thunder include Lei-kung (Lord Thunder), who is portrayed as an ugly man, with the wings of a bat, an eagle's head, and talons for hands and feet. He wears nothing but a loincloth and has several drums hanging from his waist, which he uses to make thunder. Tien Mu (Mother Lightning) holds mirrors in her hands to shine light down from the heavens. Yu-tzu (Master of Rain) sprinkles water from an urn with his sword, and Yun-t'ung (Little Boy of the Clouds) piles up storm clouds. Feng-po (the Earl of Wind) releases the wind from a goatskin bottle, and Feng-p'o-p'o (Mrs. Wind) flies among the clouds on a tiger.

A Chinese tale tells of a hunter caught in a storm. He saw that the thunder was concentrated around one tree under which a child stood holding a flag. Each time the child waved the flag, the thunder was forced to retreat. The hunter reasoned that the child was an evil spirit as the gods are unable to approach anything unclean, and the flag that the child was holding to keep the thunder at bay was made of some unclean material. The hunter loaded his gun and shot the flag out of the child's hand. As soon as he did so, the thunder spirit struck the tree with a deafening blow. The hunter, still standing nearby, felt the power of the blow and fell unconscious. When he regained consciousness, he noticed a strip of paper laid across his body. There was an inscription on the paper: "Life prolonged for twelve years for helping on the work of heaven." Beneath the still smoldering and splintered tree lay the body of a great lizard, the evil spirit in its true form.

If there is a flag of some "unclean material" holding today's thunder spirit at bay, it is that of ignorance. Despite the discrepancies between myth and science, it would be wrong to think of science as bloodless and strictly analytical, or literature void of true perception. Science and literature are not at opposite poles. Both practices require an intense curiosity and powers of keen observation. Both practices are fueled by a sense of the marvelous. As Albert Einstein put it, "The most beautiful thing we can experience is the mysterious. It is the source of all true art and science."

Samuel Colman, from *Harper's Weekly,* Oct. 19, 1878

Clouding

John P. O'Grady

The ability to recognize kinship is a god. That at least was the ancient wisdom, and it still holds true today. "I know you!" A bright idea. Love at first sight. A letter from a long-lost friend. An old photograph found in an antique shop that proves an exact likeness to your mother, your child, or yourself. Like being drawn unto like. Kinship. All of these are epiphanies, theophanies, each one tender as a bolt of lightning. Etymology suggests that the atheist is not the one who refuses belief in the divine, but the one who has been *abandoned* by the gods. The atheist is the archetypal lonely guy. I've known a few of them in my day, and I've observed that while the divine may sometimes forsake an individual, sooner or later a blessed—or terrifying—reunion takes place: the lonely guy falls in love. It may be with a person, but just as dramatically it might be with an idea, or what lies behind the idea. In any case, when the forlorn one is unexpectedly visited by a passion, life suddenly becomes head-in-the-clouds.

Traditionally, clouds are symbolic of things indeterminate. Composed of air and water, their essential nature can be attributed to neither element but arises in an obscuring of the two, a betwixt-and-between phenomenon, not unlike human beings, those nebulous creatures who themselves seem caught between realms, floating along between the shimmering horizons of birth and death, here and there, earth and heaven. Buddhist psychology refers to the aggregate of what we call personality as "the five clouds of entanglement."

But if we are clouds, we are also luminous. Xenophanes, writing at the dawn of Western philosophy, tells us that the stars are actually clouds "ignited by motion," kindled in their rising and extinguished in their setting, like coals. The sun too is a burning cloud, and as with the stars, each day it's a different cloud that is set ablaze, for no two suns are the same, though they share in the same flaring grandeur—and this goes on forever because the world is imperishable, without beginning, without end. Herein hovers a magnificent hope: entangled clouds that we are, sooner or later in our driftings we're bound to catch fire, become a star or maybe even a sun, and not just for fifteen minutes but for a whole day or night. Every soul is combustible.

In 1939 the International Commission for the Study of Clouds published a manual on how to observe these objects Shelley once called "nurslings of the sky." Before telling someone how to look for something, it is prudent to state clearly what is being looked for. Thus, the International Commission tells us, in language a little less elevated than Shelley's, that a cloud is an atmospheric event consisting of "minute particles of liquid water or ice, or of both, suspended in the free air and usually not touching the ground." It's the "not touching the ground" part that sparks my interest.

The Pythagoreans spoke of a mysterious "Counter-Earth," a sort of shadow world that occupies the orbit exactly opposite to the Earth, so it's always behind the sun, hidden from our view. I like to imagine this Counter-Earth as a place where each of us has a provocative counter-self who lives a colorful counter-life amid a vibrant counter-culture, an existence that is not really opposite to the one we enjoy or suffer here, but rather is entirely *other* to it, much as our dream lives are not opposite but *other* to our waking lives. Though our memory of it may be sketchy, we visit this Counter-Earth every night. Perhaps clouds are its allegory, opening a correspondence between the two worlds. In this sense, each cloud ought to be received like a love letter—or a ransom note—from beyond, endlessly unfolding across the sky. Another cloud book that recently came into my hands—B. J. Mason's *Clouds, Rain and Rainmaking*—cautions its reader: "At first glance a cloudy sky may appear chaotic, but the perceptive observer will discern some semblance of order, the existence of recognizable patterns." Things up there in the sky are not what they seem, say the scientists, and this resonates nicely with what various occult traditions have been fond of saying for millennia: "As above, so below." It is regrettable that the International Commission's manual offers no additional speculations in this regard.

Knowledgeable authorities in the past, however, were less hesitant to enter these darkling realms. The druids, for instance, are reported to have practiced a form of cloud divination. When a king or queen wanted a glimpse of the future, the druid was dispatched to the summit of a nearby hill or mountain to consult the clouds, much as the augurs in ancient Rome gleaned insight by watching the flight of birds. Before that was Moses, who climbed to the top of a mountain to converse with God, God who would meet him only under the cover of a thick cloud: "And the glory of the Lord rested on Mount Sinai, and the cloud covered it for six days." And coming down to us from fourteenth-century England is a book of mystical instruction titled *The Cloud of Unknowing*. I like to think of it as a manual on how to observe the weather of the mind. "When I say darkness," the anonymous author explains, "I mean thereby a lack of knowing. . . . And for this reason it is not called a cloud of the air, but a cloud of unknowing, that is between thee and thy God." He also offers this salty advice to those who would plunge into such obscurities: "Short prayer pierces heaven."

We human beings remain fascinated by clouds, perhaps even more so today than in earlier times, because in their shape-shifting inexactitude, their openness to the world, clouds seem so entirely *other* to our rock-solid world of property rights, scientific and historical "facts," fixed identities and the politics that go with them. "I'm a Republican!" "I'm a feminist!" "I'm an environmentalist!" "I'm an antiglobalization anarchist!" What's in a name? Wrong question. Better to ask, What's outside a name? What you'll find there is nothing but clouds, free and easy wandering. One of my favorite Zen koans contains the line, "I am not a human being!"—a gentle reminder of the clouds whereof each of us is composed.

Nevertheless, we persevere in our efforts to pierce heaven, now preferring telescopes to prayers. Contemporary cosmologists would lead us by the eye down their long tunnel and through their thick lens to the very heart of the universe, which is revealed not as the still point of old, but as a noisy "Big Bang." Contemplation replaced by a fireworks display. Yet we are estranged from our very selves. Who among us can penetrate even the little secret of our own shifting moods, those storms of passion that characterize our most ordinary affections? "From day to day," writes Emerson, "the capital facts of human life are hidden from our eyes. Suddenly the mist rolls up, and reveals them, and we think how much good time is gone, that might have been saved, had any hint of these things been shown."

Consider that holy logic chopper Thomas Aquinas, who devoted his life to composing his multivolume magnum opus, the *Summa Theologica,* regarded by many as the most important of all Christian theological works. It garnered for him a well-deserved renown, but no entry into that heaven he so diligently stormed with his intellect; instead, attainment came near the end, unexpectedly and without effort. The story goes that just a few months before his death, Aquinas was celebrating a mass when suddenly he had a mystical experience. After that he gave up any further work on his *Summa;* he quit writing altogether. When somebody asked him why, he replied: "All that I have written seems to me like straw compared with what has now been revealed to me." It's often the case that what we so desperately long for is nearer to the heart than the heart is to itself.

When I was in graduate school studying literature, I came up with an idea that, had I pursued it, would have made me a rich man. Instead, I spent all my time on poetry, which is why now I write essays like this just to pay for the groceries. But anyway, it occurred to me that given American consumers' insatiable desire to lay claim upon the intangible—everything from acquiring a stock portfolio to owning a "piece of the Rock"—why not give them the ultimate pie-in-the-sky delusion: Why not offer them an opportunity to buy and sell *clouds*? Allow me to explain.

At the time, I was living in Maine. My plan was to wait for one of those bright August afternoons that occur on days after a powerful cold front has moved through, a day when the sky is filled with those fluffy cumulus clouds, the sort Daisy Faye wanted to push Jay Gatsby around in, clouds clearly detached from one another and sharply delineated, each insisting upon its own individuality (if only for a moment), the kind that John Muir lamented were "hopelessly unsketchable and untellable."

But not un-photograph-able.

That was my plan—to take pictures of individual clouds, and for each one print up an elegantly lettered deed of title on parchment, then put them up for sale. A typical deed would read something like this: "Witnesseth, that in consideration of Ten Dollars ($10.00), in hand paid, the receipt whereof is hereby acknowledged, the said grantor does hereby grant and convey, sell and confirm unto said grantee, his heirs and assigns, all that certain piece or parcel of cloud situated, at 1:18 P.M. on the Eleventh of August, Year of Our Lord 1983, in the sky above Pemetic Mountain in the County of Hancock, State of Maine." One caveat, one encumbrance let us say, on this transaction: while each deed would confer full title to a particular cloud and specify the exact time and location it was last seen, the burden of again locating said cloud rests entirely upon the buyer. That's why every deed would be accompanied by a photograph of the cloud—at least then the buyer would have a clear image of what it is he's looking for and a place to start. Thus, he's already two steps ahead of most idealists.

Consider this cloud scheme my contribution toward keeping the nation's economy on track. We all want to own a piece of the American Dream, but given a country with an ever-increasing population and a "limited resource" of land, comparatively few will be able to own real estate, but everyone can have title to *un*-real estate—we can all be cloud owners! Poverty, says Plato, is not a function of small property but of immense desires. So give them clouds! People could even assign names to their cloud-property, much as the wealthy do for their estates: "Sunnybank," "Olana," "Onteora." The sky's the limit when it comes to the number of clouds that can be put on the market. They are not, at least in many parts of the globe, a limited resource, they can't be used up, and the world keeps making more. In this sense, clouds are a lot like kisses: you can keep giving them away but they never run out.

My plan was to open a little Un-Real Estate office down on Cottage Street in Bar Harbor. "Cloud Nine" would be its name. Instead of pictures of houses posted in my storefront window, there would be clouds. Had my life not taken a very different course, and had the success of this business been anywhere near commensurate with my dreams for it, franchises would have spread out quickly, and today you'd be looking at a Cloud Nine Un-Real Estate office there in your town. Instead, you have Wal-Mart and an awful lot of undeeded property floating by in the sky.

Alas, I wasn't the first to come up with the idea of capturing clouds on film. Late in his life, the photographer Alfred Stieglitz started making photographs of clouds. By all accounts, his illustrious career seemed to be winding down by 1915, but then something happened to reawaken his creativity. These days art historians argue that this rejuvenation was a result of aesthetic insight gleaned from the work and ideas of Picasso, the cubists, and other modern European artists. But I'm guessing it had more to do with Georgia O'Keeffe. The two met in 1916, and not long after that they became lovers. "She is much more extraordinary than even I had believed," Stieglitz wrote, in his modest way, to a friend in 1918. "In fact I don't believe there ever has been anything like her—Mind and feeling very clear—spontaneous—& uncannily beautiful—absolutely living every pulse beat." These are the words of a man whose mind is socked in with the clouds of Cupid. Stieglitz, no doubt, had that ability to recognize kinship—even in the fog—and he acted on it, though it cost him a marriage of twenty-five years.

Over the next decade, he generated an extended serial portrait of O'Keeffe. Many of these images, which seem almost Tantric in intention, depict O'Keeffe in various stages of undress as well as fully nude. In these photographs, Stieglitz, who was fond of saying that he made love through his camera, was not only challenging the taboos of his day against public displays of sexuality, but also intimating dark possibilities of nature-magic and rank fertility. After all, he had been reading Freud, who proclaimed the human body is on the one hand sacred, consecrated, and a veritable temple, while on the other "uncanny, dangerous, forbidden, and unclean." Talk about a Counter-Earth! O'Keeffe's lean and taut body affected the aging Stieglitz mightily. "When I look at her," he confesses in another letter, "I feel like a criminal.—I with my rickety old carcass [he was a quarter-century older than O'Keeffe] & my spirit being tried beyond words."

It was right on the heels of making these erotically charged photographs of O'Keeffe (not to mention other women) that Stieglitz turned his peculiar form of lovemaking toward the clouds. The same creative energy (call it libido) that he had lavished upon making photographs of the female body was now released skyward, like a bunch of doves. In a 1923 letter describing his new work to the novelist Sherwood Anderson, Stieglitz comes off sounding like a lickerish adolescent, admitting almost boastfully that "after many days of passionate working—Clouding!—I stopped. I had to catch my breath."

In all fairness to him, it should be remembered that Stieglitz *always* talked this way when it came to his art. Late in life, looking back on a long career, he reflected: "It's difficult to understand today the passion and intensity I poured into Photography in those early years. I spent hours, days, weeks and months. Photography had become a matter of life and death." Freud called this kind of thing sublimation, "the instinct's directing itself towards an aim other than, and remote from, that of sexual gratification." Perhaps, but such claims are always dubious. The psychologist's words smack of a clinical interiority, just a little too walled in from the vast Outside that was the subject of Stieglitz's art. More in the spirit of these photographs, it might be said that the sky was the only place ample enough to contain a vision as grandiose, and as lonesome, as his; after all, he was an artist, one of those personalities, to borrow the words of the International Commission for the

Study of Clouds, "suspended in the free air and usually not touching the ground." Or, as Horace puts it, "I shall not die, my sublimations will exalt me to the stars."

Setting his own sights a bit closer to the earth was the great artist-ornithologist John J. Audubon, who upon arriving in the United States in the early nineteenth century felt "prompted by an innate desire to acquire a thorough knowledge of the birds of this happy country." Dressed in billowing satin breeches and fine silk stockings, he stalked the fields and woods of America, trusty rifle by his side, enthusiastically "collecting" birds and other wildlife so that he might sketch them in lifelike detail. In later years, he provided an account of his artistic technique, which first came to him in a dream. It involved inserting metal wire into the lifeless bodies of birds he shot in the field, then setting up the feathery mannequins and manipulating them into realistic poses. "Reader," he says, fondly recalling the kingfisher that opened the door to his success, "this is what I Shall ever call my first attempt at Drawing actually from Nature, for then Even the eye of the Kings fisher was as if full of Life before me whenever I pressed its Lids aside with a finger.—" After this, his passion blazed forth into professional ambition, and, using the same method of shooting and wiring, he set himself to the task of depicting every bird in America, the results of which are still to be marveled at in his massive books.

Perhaps most poignant in all the work of John J. Audubon is his description of the passenger pigeon, once the most numerous bird on earth. Flocks—or should we say dense clouds—of these birds were reported over a mile wide and, by some estimates, over three hundred miles long. They numbered in the billions. Remembering the autumn of 1813, Audubon writes: "The air was literally filled with Pigeons; the light of noon-day was obscured as by an eclipse." He goes on to recount hordes of people out in the Kentucky woods, with horses and wagons, guns and ammunition, there to shoot pigeons. In just a few hours, tens of thousands of them were slaughtered. Some of the dead birds were gathered for the market, but most were left to lie on the forest floor, where hogs were turned loose to feed upon the carcasses. What the hogs missed, bears, raccoons, possums, and vultures scavenged. Anticipating readers' distress at the news of this bloody havoc, Audubon assures us that nothing, barring the elimination of the forests, could do any harm to a species as numerous and as fertile as the passenger pigeon. Alas, he proved less able a prophet than a painter. The last passenger pigeon, named Martha, died alone in captivity at the Cincinnati Zoo on September 1, 1914. Now when we read

Audubon's words, not a bird but a ghost is evoked: "When an individual is seen gliding through the woods and close to the observer, it passes like a thought, and on trying to see it again, the eye searches in vain; the bird is gone."

One day toward the end of his life, after gazing on the mists rising from the Hudson River, Audubon turned to his easel to begin painting. A curious thing happened. The morning sun failed to burn away the low-lying clouds, which indeed seemed to be thickening around his canvas. Yet when he looked up, the old man discovered everything at a distance was bright and clear. The mist was not in the atmosphere but in his eyes: his vision was failing. When this phenomenon persisted over the next few days, Audubon realized that his career as a painter was over. During the next few years, perhaps as the result of a stroke, dementia set in. As one old friend put it: "The outlines of his beautiful face and form are there, but his noble mind is all in ruins. It is indescribably sad." Audubon withdrew further and further into the dim recesses of himself, until he no longer recognized anyone around him. His kinship now was exclusively with the dark.

Perhaps this is the way of all true artists. They're like the old druids and Moses and maybe anybody who is madly, passionately in love: poking around up there in the sky, looking for something beyond reach, calling out to it and hoping it responds, so that, should it draw near, they can pounce on it. There's a certain impossibility in all these endeavors—to fix the future, to behold a god in all its splendor, to capture nature in its bare reality, to embrace the loved one forever. Isn't it the case that all who reach for the sky are bound to come home empty handed? "What can poor mortals say about clouds?" lamented John Muir.

Dorothy Norman, long-time assistant to Stieglitz, reports that not long before he died, Stieglitz was asked, What is the perfect photograph? He responded by spinning a fantasy in which he himself had just taken that perfect photograph and was now holding the glass negative in his hands, reveling in his accomplishment. "It is exactly what I wanted!" the artist exclaims. But suddenly the glass slips from his hands and shatters on the floor. "I will be dead," Stieglitz concludes sadly, "and no one will ever have seen the picture nor know what it was." The moral of the story seems to be that there is no place for perfection in this world, a view consonant with Plato's famous assertion that no sensible person would try to express his or her grandest thoughts—those thunderheads of the mind—in a form that is unchangeable.

In this regard too I think of John J. Audubon, blasting away with his rifle at his beloved American birds soaring in the limitless sky, that he might bag a few ideal examples of each species and immortalize them in his art. Yet when all the birds are shot and all the drawings complete, each specimen splendidly depicted across the immense pages of his glorious books, did old John J., suffering the dementia of his last years, "his mind all in ruins," cast his hazy marksman-artist eye upward through the bird-free sky and gaze longingly upon those high and companionable clouds, the only things now remaining between himself and heaven?

Joachim d'Alence, frontispiece to *Traittez de barometres, thermometres, et notiometres, ou hygrometres,* 1688. Courtesy of the NOAA Collection

From *Oxygen*

Carl Djerassi and Roald Hoffmann

Cast of Characters

Antoine Laurent Lavoisier, 34 years old. (*French chemist, tax collector, economist, and public servant; discovered oxygen*).

Marie Anne Pierrette Paulze Lavoisier, 19 years old. (*Wife of the above*).

Joseph Priestley, 44 years old. (*English minister and chemist; discovered oxygen*).

Mary Priestley, 35 years old. (*Wife of the above*).

Carl Wilhelm Scheele, 35 years old. (*Swedish apothecary; discovered oxygen*).

Sara Margaretha Pohl (Fru Pohl), 26 years old. (*Became Mrs. Scheele three days prior to Carl Wilhelm's death*).

Court Herald (*off-stage voice*).

ACT 2

Scene 14. *Stockholm 1777. Suggestion of a palace setting. At center is a demonstration table. Stage right is a lectern. Upstage right is a free-standing screen, which could be used for shadow play. Actual and simulated experiments will be done at this table; projections may be shown on the side screens. Left stage are three chairs for the women.*

Court Herald's Voice: (*Forceful, somewhat pompous voice*) Your Majesties, esteemed guests! Throughout Europe, pneumatic chemistry is in the air. A dispute has arisen: Who, among these great savants, discovered the vital air supporting life? (*Pause*) A golden medal . . . with the likeness of our King Gustavus III . . . will be struck in honor of the true discoverer. And our King is famed for his munificence in other ways.

Priestley: *(Aside)* As he squanders the people's money . . . *(Trumpets)*

Court Herald's Voice: Let the Judgment of Stockholm begin! And let the three savants be their own judges! Vital air! *(Pause)* Who made it first?

Scheele: *(Quietly, but quickly, beating the other men to the draw)* I did. And called it *eldsluft* . . . a good Swedish word for fire air.

Priestley: But is that not air deprived of all phlogiston? The air that inflames all things? That is why I named it "dephlogisticated air." *(Pause)* But dear Scheele . . . where should we have learned of your discovery?

Scheele: In my book, about to appear . . .

Priestley: I made that air by heating *mercurius calcinatus* in 1774 and . . . *(Raises voice for emphasis, while addressing Scheele)* communicated that discovery in the same year!

Lavoisier: *(Smiling)* Mes amis! He who starts the hare, does not always catch it.

Scheele: There is no hare to catch if someone does not start the hunt!

Lavoisier: It is we who must decide who first captured the essence of that vital air. . . .

Priestley: *(Sarcastic)* And what does that mean?

Scheele: It is essential to know who made the air first . . .

Priestley: . . . for it is the invention that will be remembered by posterity, not its ephemeral interpretation . . .

Lavoisier: *(Shifting the subject)* Let us do the experiments we judge vital in this matter. Whose experiment will come first?

Scheele: Monsieur Lavoisier, do me the honor of performing the experiment I brought to your attention some three years ago in my letter—

Lavoisier: I know of no letter—

Scheele: *(Gets letter from Fru Pohl)* Let me read it for you.

(Lights DIM; *spots on two men. This is the first of three experimental scenes. The stage is*

darkened, except for spots on the bench and on the man who performs the experiment, as well as the one who directs him.)

Scheele: Dissolve silver in acid of nitre and precipitate it with alkali of tartar. Wash the precipitate, dry it, and reduce it by means of the burning lens in your apparatus.

Lavoisier: But I brought no burning lens!

Scheele: My apologies. With the burning lens at hand, Monsieur. When I first wrote, I thought of your celebrated giant lens, which was so much superior to what was available in my pharmacy. No matter. A mixture of two airs will be emitted. And pure silver left behind.

Lavoisier: And then?

(Light DOWN *on men, who continue their experiment, possibly in mime. Light* UP *on women stage left.)*

Fru Pohl: Apothecary Scheele once invited me into his shed. To show me an experiment he had done earlier in Uppsala. He was bubbling the newly formed fire air through a kind of water.

Mme. Lavoisier: It must have been limewater.

Mrs. Priestley: It turned cloudy, didn't it?

Fru Pohl: How do you know?

Mrs. Priestley: I've listened to Joseph's lectures on fixed air.

Mme. Lavoisier: The same air we expire . . . the one we remove by passage through limewater.

Fru Pohl: In the remaining air, he bid me thrust a splint that had blown out. Just a glow of a coal at its end. It was toward evening.

Mrs. Priestley: And it flared in brightest flame . . . and kept burning!

(The flaring up of the splint in the men's experiment coincides with its mention by Mrs. Priestley. Lights OUT *on women,* UP *on men)*

Scheele: I did that experiment in 1771, three years before your experiment, Dr. Priestley, in a pharmacy in Uppsala…with equipment much more modest than now put at our disposal by your Majesty.

Priestley: Yet you did not report it?

Scheele: I told Professor Bergman . . . I thought he would tell others. I had to earn my wages. I wanted to experiment. I had but little time to write of my observations. (*Waves page*)

Priestley: Your experiment was with a silver salt.

Scheele: I obtained the air over the next three years in many different ways. Including red *mercurius calcinatus*, as you did.

Lavoisier: That red mercury compound—it is also how we . . . Dr. Priestley and I . . . made that air.

Priestley: <u>We</u>? (*Pause*) We were not in the same laboratory, Monsieur Lavoisier! Pray speak clearly of who did what and when. (*More heated*) More than once, my experiments in pneumatic chemistry were cited by you—

Lavoisier: Is that a reason to complain?

Priestley: Only to be then diluted…if not evaporated.

Lavoisier: How did I do so?

Priestley: You write (*Heavy sarcasm*), "<u>We</u> did this . . . and <u>we</u> found that." Your royal "we," sir, makes <u>my</u> contributions disappear . . . poof . . . into thin air! (*Pause*) When I publish, I say, "<u>I</u> did . . . <u>I</u> found . . . <u>I</u> observed." I do not hide behind a "<u>we</u>."

Lavoisiser: Enough of word plays. (*Louder*) What now?

Priestley: <u>I</u> made that air first . . . and did so alone. And I will now show you how I accomplished that. Mr. Scheele, will you perform the experiment?

Scheele: It will be an honor to do so.

(*Both men step to demonstration table; lights* DIM)

Priestley: In August of 1774, I exposed *mercurius calcinatus* . . . the red crust that forms as mercury is heated in air . . . in my laboratory to the light of my burning lens. As the red solid is heated, an air will be emitted, while dark mercury globules will condense on the walls of the vessel. You will collect the air by bubbling it through water. As soon as the gas appears, be careful, Mr. Scheele, to catch it under water.

Lavoisier: But where is your balance, Dr. Priestley? Shall the gas not be weighed?

Priestley: A timepiece is sufficient. We have here two chambers . . . one with ordinary air . . . the other with my new dephlogisticated one. Mr. Scheele, now take a mouse . . .

(*Lights* DOWN *on the men, who continue the experiment with two jars and two mechanical mice in a small cage. Lights* UP *on women.*)

Mrs. Priestley: I asked him—why mice?

Fru Pohl: And?

Mrs. Priestley: And he said: Mice live as we do. Would you use English children?

Mme. Lavoisier: They live on a part of ordinary air.

Mrs. Priestley: Then he placed one mouse in a jar of plain air.

(*In the darkness, Scheele can pretend allowing the mechanical mouse to escape and then retrieving it by its tail.*)

Fru Pohl: Where it died in time.

Mrs. Priestley: How do you know that?

Fru Pohl: Apothecary Scheele showed me.

Mme. Lavoisier: It is a well-known fact, described also by other savants.

Mrs. Priestley: And then he placed the other one in—

(*Scheele mimics this experiment with the second mouse.*)

Fru Pohl: Fire air . . .

Mrs. Priestley: My Joseph's dephlogisticated air . . .

Mme. Lavoisier: And it lived much longer, did it not? This is why we call that new air "eminently respirable" or vital air.

Fru Pohl: (*Laughs*) With living things, Carl Wilhelm can be clumsy. He often dropped them! But we know mice in the country. If I didn't catch them, the cats did.

Mrs. Priestley: I detest mice.

(*Lights* OUT *on the women,* UP *on the men*)

Lavoisier: There is no doubt that Dr. Priestley's method produces vital air. But—

Mrs. Priestley: <u>But</u>, Monsieur?

Lavoisier: Now is my turn. May I proceed?

Scheele, Priestley: Of course.

Lavoisier: We just observed a mouse living longer in the vital air we have all made. Yet in the end that mouse also dies, as the air is depleted. However, in my <u>own</u> work . . . I have moved far, far beyond watching mice die. Your Majesty, gentlemen! This air . . . that I propose we henceforth call *oxygène*—

Priestley: *(Interrupts)* I object, sir! It's easy to call something by a new name . . . when you don't know what you have! Be descriptive, sir! Why not dephlogisticated—

Lavoisier: I know the air as well as you do, Monsieur. "Oxy" is Greek . . . for acid. And since I believe our air to be associated with all acids, I am being descriptive . . .

Priestley: Descriptive? Bah! You, sir, are being acid . . . but our dephlogisticated air is not.

Lavoisier: Allow me the courtesy to continue. This air is at the heart of all chemistry. I have shown that when we breathe, the wondrous machinery of the body transforms a given weight of *oxygène* . . . into other gases and water.

Priestley: But that is obvious!

Lavoisier: Not until you weigh it! For that . . . *(Confronts Priestley)* . . . a timepiece is <u>not</u> sufficient. . . . Since nothing is gained . . . nor lost in this world . . . be it in the economy of a country or a chemical reaction . . . the balance sheet of life's chemistry must be determined.

Priestley: *(Dismissive)* The *banquier* still counting his money. . . .

Lavoisier: *(Ignores Priestley's comment)* I have brought from Paris a suit of rubber I have devised. It catches all the effluents of the body . . . to show us that the equation balances. *(Pause)* Dr. Priestley, are you prepared to perform the experiment?

Priestley: Indeed I am . . . even weighing things on your balances.

Lavoisier: My experiments are quite complex technically. Perhaps Mr. Scheele will help you? *(Scheele joins Priestley)*

Priestley: It appears we require a volunteer for the experiment . . . to wear your modern suit of armor. (*Looks around, turns to his wife*) Mary?

Mrs. Priestley: (*Most reluctant*) I would help you, Joseph, but I fear for my life in that French contraption.

Priestley: Don't be afraid. It's only science.

Mme. Lavoisier: I will do it!

(*Mme. Lavoisier marches up, with determination. She picks up "rubber suit," not unlike old-fashioned diving or scuba suit. Scheele and Priestley help her put it on, behind the screen. Projection of one of her drawings of the experiment may appear.*)

Lavoisier: Not only must you weigh my spouse . . . you must also weigh her suit. The measurements will take several hours.

Mrs. Priestley: (*Shocked*) Poor Madame!

Lavoisier: Quantitative experiment is a hard mistress.

(*Lights* DIM *on men, remain on Mme. Lavoisier in suit, and* UP *on Mrs. Priestley and Fru Pohl.*)

Mrs. Priestley: She sketched her husband's experiment.

(*Projection of one of Mme. Lavoisier's drawings of the experiment appears on screen for remainder of their conversation.*)

Fru Pohl: But why? For her pleasure?

Mrs. Priestley: As a record, I suppose.

Fru Pohl: But why should a "record" be needed?

Mrs. Priestley: To give others evidence of what was done, of course.

Fru Pohl: And when it was done, I should think.

(*Lights* DIM *on women*)

Lavoisier: (*Addresses Priestley*) I trust you took care . . . for the margin of error must not be more than 18 grains in 125 pounds. What do you find?

Priestley: Mme. Lavoisier has lost some weight . . . (*Mme. Lavoisier seems weak, but smiles*) but when we take into account the water and fixed air breathed out, there's indeed a rough balance.

Lavoisier: Now we have verified each other's experiments. Now that we see that nothing is created—

Priestley: Except by God.

Lavoisier: Nor lost.

Scheele: Except by Man.

Mme. Lavoisier: Or woman. Especially when she is the subject of an experiment.

Lavoisier: (*Driving his point home, and refusing to enter the banter*) Gentlemen! That crucial mass balance (*with emphasis*) . . . punctures phlogiston's balloon.

Priestley: Not in my view, sir! (*Addresses Lavoisier*) The experiment you so laboriously had us do, with balances, and the patient suffering of your wife, did demonstrate . . . I readily confess . . . <u>one</u> function of your (*assumes sarcastic tone*) "eminently breathable air." (*Pause*) But, Monsieur, you did not show us <u>how</u> you made that air.

Lavoisier: I knew my air was there in ordinary air. . . . Did I not see metals combine with it . . . with sulfur . . . or with phosphorus?

Priestley: That does not tell us how you produced the dephlogisticated air. . .

Lavoisier: Pray stop calling it "dephlogisticated," Dr. Priestley. The name derives from a theory that is *passé.*

Priestley: Not for me.

Scheele: Nor for me.

Lavoisier: Why not a new name for the air, to avoid this argument?

Priestley: Call it (*dismissively overemphasizes French pronunciation*) *oxygène?* And yield to the tyranny of a nomenclature, invented by you?

Lavoisier: (*Angry*) When a new structure is needed for a science . . . when, indeed, there must be a revolution, new names are also required.

Priestley: But you did not know what that gas was!

Lavoisier: I saw the need for one air explaining rusting, burning, and respiration!

Priestley: (*Heatedly*) But until that October dinner in Paris when I informed you of my observations . . .you did not know the nature of that air . . .

Scheele: (*Atypically forceful*) And until that October day when you got my letter which told you how to make fire air. . . .

(*They argue simultaneously to the end of the scene.*)

Lavoisier: I had begun my experiments with *mercurius calcinatus* . . .

Priestley: Only after you heard of what I discovered . . .

Scheele: You did not know how to make that air . . .

Court Herald's Voice: (*Sound of tapping staff*) Order! Order! Gentlemen . . . his Majesty is vexed. (*Pause*) Royal displeasure is the only judgment you will receive today!

(*Lights* FADE)

End of Scene

The Subtle Humour

Bruce F. Murphy

Such as is the air, such be our spirits; and as our spirits, such are our humours.
—Robert Burton, *The Anatomy of Melancholy: what it is, with all the kinds, causes, symptomes, prognostickes and severall cures of it*

Imaginary air, specifically, is the hormone that allows us to grow psychically.
—Gaston Bachelard, *Air and Dreams*

These two aspiring thinkers—philosophers of air—confront each other across four centuries, as through a looking glass. From opposite ends of the Enlightenment, they challenge the very definition of air. Both statements above, written 350 years apart, attribute our health and well-being to air. On the face of it, that isn't enough to make them interesting; to modern society, the importance of air is sufficiently uncontroversial. Isn't air a composite gas made up of oxygen, nitrogen, and various other gases, pollutants, and particles? Haven't we had a Clean Air Act and conferences on global warming? But Burton and Bachelard are operating in a more rarefied atmosphere. Burton's claim that air determines our spirits and our "humours" and Bachelard's notion that "imaginary air" is a kind of psychic growth hormone both seem quite preposterous. Air is, after all, so ordinary. Most people seem content knowing that air flows in and out of us, and if the flow stops for any length of time, we die. We understand that we must breath to survive, without feeling the need to think about air. But for Burton and Bachelard, thinking about air is a path toward understanding the universe and a stimulus to the imagination. Air is the most inescapable of the four classical elements, and therefore the most ethereal. It is, in a sense, the realm of thought, of imagination. To refuse it is to be imprisoned in the coarser elements.

To begin cultivating a philosophy of air, one must first understand the pull of unimaginative thinking. Burton and Bachelard were both renegades within the bastions of conventional wisdom. Burton was an Oxford divine of the seventeenth century who went outside the bounds of theological orthodoxy because he was fascinated by the new science. Bachelard, a former mailman who became a professor of the philosophy of science at the Sorbonne, revolted against twentieth-century positivism because he was fascinated by "the material imagination"—an idea most of his colleagues would greet as a contradiction in terms. Both were imaginative men who understood the orthodoxies of their time to be earthbound and who identified air and flight with imagination.

If we liberate ourselves from scientific thinking, we can recuperate the figurative value of air—a shift from respiration to aspiration, mere existence to living. Bachelard rejects the rationalistic-scientific demotion of the imagination in favor of reason, "as though the imagination were unproductive time-off from a persistent, affective occupation." Bachelard's recovery of the value of imagination hinges on the idea that we live not only in the world of our perceptions but of our imaginings. In fact, the imagination alters perception: "We always think of the imagination as the faculty that *forms* images. On the contrary, it *deforms* what we perceive; it is, above all, the faculty that frees us from immediate images and *changes* them." Without imagination, "there is only perception, the memory of perception . . . an habitual way of viewing form and color"—and habit, he says, "is the inertia of psychic development" and "the exact antithesis of the creative imagination." But imagination thus has no value in Bachelard's world of rationalism, work, and money. It doesn't "do" anything. We live in an explained universe (even if all the particulars are not understood, the scientific paradigm is firmly in place), to which any imaginative "additions" are fated to read as falsifications, delusions, or fantasies.

Burton lived in an equally deadened time, but it was the deadness of another kind of certainty, that of theology. Ostensibly an investigation into the causes and cures of melancholy, or madness, Burton's *Anatomy of Melancholy* is an enormous compendium of human theories about the world. For the most part, Burton enjoyed sifting through twenty centuries of learning without conducting any original research, content to observe that apparently melancholy can be caused by anything and experienced by anyone. But he also recognized that traditional knowledge was wildly contradictory and appeared to be great graveyard ideas, some of which, at the very least, must be inaccurate.

For several thousand years, the theory that the world was composed of the four elements of air, earth, fire, and water and that the planet earth was the center of the universe around which all revolved had been accepted. In the Christian era, it had been officially sanctioned, and during Burton's own lifetime, people had been burned at the stake for daring to think otherwise. But by observation, Kepler, Copernicus, and Galileo had done much to smash the beautifully balanced yet unreal Ptolemaic model of the universe. The new discoveries and the questions they brought up fascinated Burton. In his "Digression of Air," Burton gives his proto-Enlightenment curiosity full rein. He takes off, literally, into the air. It is in the air that he searches for answers. Burton's digression has been called the first essay in climatology; at the same time, it illustrates Bachelard's principles of creative imagination and deformation of the "real"—except that Burton approaches the matter from the other end, so to speak. He lived at a time when perceptions of reality were all severely deformed, in the scientific sense. One could actually believe that "the air is not so full of flies in summer as it is at all times of invisible devils," as Paracelsus did; that one risked inhaling devils with every breath; and that they "go in and out of our bodies, as bees do in a hive, and so provoke and tempt us as they perceive our temperature inclined of itself."

Burton's digression is largely composed of questions. Some are meteorological, others climatic, still others zoological or ethnological. The new scientific spirit that was in the air in Burton's time had begun to tear apart the classical universe simply by taking a closer look at it. The classical model had stressed stasis, as befitted a universe constructed on a divine plan by an infallible intelligence. But on examination, the new scientists had found the universe complex, lopsided, and puzzling. Observation raised more questions than it solved. But what especially comes through in the "Digression of Air" is the joy of recognizing the natural world for the first time: "Whence comes this variety of complexions, colours, plants, birds, beasts, metals, peculiar almost to every place? Why so many thousands strange birds and beasts proper to America alone." (Burton rightly asks that if they all did come out of Noah's ark, why are they not by now dispersed generally?) Disparities among people suddenly become interesting rather than outlandish or repulsive. Burton wonders "whence proceed that variety of manners, and a distinct character (as it were) to several nations? . . . Is it from the air, from the soil, influence of stars, or some other secret cause?" He wants to know where the water goes that pours from rivers into the oceans ("how is this water consumed? by the sun or otherwise?"), and what happens to birds when they disappear in winter ("whence they come, whither they go, *incompertum adhunc,* as yet we know not"). And of course he wants to know about air—not only why it varies from place to place, but why it is different in places that should be the same: "But this diversity of air, in places equally site[d], elevated, and distant from the Pole, can hardly be satisfied with that diversity of plants, birds, beasts which is so familiar to us." Burton wants to understand the biosphere from top to bottom.

The real focus, however, is astronomical, which tests the limits of the air. It is hard to imagine, but true, that the debate over what air *was* could be literally life threatening. Why, we wonder, was air so important?

As though in a preview of the conflict of the gradualists and catastrophists of the nineteenth century, Burton says, "In all other things nature is equal, proportionable, and constant . . . why are the heavens so irregular?" The scholastic mind was disturbed most deeply by "this rash placing of stars" in the heavens, which were, after all, supposed to be the heavens. And yet certain areas of the sky were blank and others cluttered; Galileo "by his glasses" had observed the Milky Way and found "that *via lactea* a confused light of small stars, like so many nails in a door." The classical model seemed quite exploded, but Burton is a man of his time, and he still wants to rectify the new learning with the old. He wants, for example, to know what the moon is for. He also wants to know what lies between us and it, and that involves the limits of the atmosphere—or, as he would have said, the air:

> P. Nonius Saluciensis and Kepler take upon them to demonstrate that no meteors, clouds, fogs, vapours, arise higher than fifty or eighty miles, and all the rest to be purer air or element of fire: which Cardan, Tycho, and John Pena manifestly confute by refractions, and many other arguments, there is no such element of fire at all. If, as Tycho proves, the moon be distant from us fifty and sixty semi-diameters of the earth, and, as Peter Nonius will have it, the air be so angust, what proportion is there betwixt the other three elements and it? To what use serves it? Is it full of spirits which inhabit it, as the Paracelsians and Platonists hold, the higher the more noble, full of birds, or a mere vacuum to no purpose?

Whether nature abhors a vacuum, Burton certainly did. He couldn't conceive of the idea that between us and the moon there was simply nothing. The theory of the four elements did not allow for that; it would have required the creation of a fifth element or nonelement—the perfect absence of the other four. Are the air and the matter of the heavens "two distinct essences"? No one was ready to admit that. So if space wasn't made of earth or water or fire, then it had to be air. Burton quotes several "late mathematicians," including Giordano Bruno, as theorizing that space is occupied by air, "saving that the higher still the purer it is, and more subtle"—on the analogy of the thinness of the air on a mountaintop. The air must become "angust," or rarified.

In addition to the problem of "the essence and matter of the heavens," Burton takes up the vexatious and equally airy "main paradox of the earth's motion, now so much in question." That this learned but unworldly vicar could even consider some of the new hypotheses that had been tabled about the motion of the earth (which, he notes, the Church of Rome had condemned as heretical in 1616) is a testament to his imaginative freedom. He was able not only to observe the world but to marvel at it; the great flight of his "Digression of Air" allows us to experience a moment of primary wonder (or creative imagination). Rather than reject the displacement of the earth as the center of the universe as offensive to God, he asks, "Why should not an infinite cause (as God is) produce infinite effects?" The vast expanse of empty air in which the earth supposedly floated dazzled him. He even dreams, as Bruno did in *De l'infinito, universo e mondi* (1584), of infinite worlds: "If our world be small in respect, why may we not suppose a plurality of worlds, those infinite stars visible in the firmament to be so many suns, with particular fixed centres; to have likewise their subordinate planets, as the sun hath his dancing still round him?"

To realize that Burton's flight of fancy in the "Digression of Air" led into dangerous regions, and that air was a concept bitterly to be fought over, we need only to remember that the "late mathematician" Bruno, a great proponent of the new physical and astronomical theories, was burned at the stake as a heretic in 1600, at which point Burton had already been at Oxford for seven years.

Perhaps that is why Burton's creative imagining of the possibilities of the air and space comes to an end. He shrinks, finally, from some of the grander possibilities he has conjured up. The most liberal answer Burton could give was "*Difficile est nodum hunc expedire, eo quod nondum omnia quae huc pertinent explorata habemus*—This difficulty is hard to resolve, because we do not yet possess all the required data."

Bachelard is a Burtonian figure and engages in similar flights of fancy. Although a philosopher of science, he took up a number of subjects that would seem to be nonstarters from a scientific perspective, such as a "poetics of space," a "psychoanalysis of fire," and a "physics of poetry." Against bellicose rationalism and self-satisfied scientism, Bachelard attempts to give poetry, imagination, and intuition what he deemed their proper weight, just as Burton braved the stake and gave weight to the evidence of a new science. Barchelard tried to keep an open mind in which neither mode of thought, rational or imaginative, was outlawed. He asserted the importance of the imagination, finding, "If we cannot imagine, we cannot foresee," in which case we can no longer act but merely react.

Bachelard wrote four books dealing with the "material imagination"—one each for earth, water, fire, and air. "I think I am justified," he writes in *Air and Dreams,* "in characterizing the four elements as the hormones of the imagination." It is not hard to see how the "hormone" of air, as he called it, fits into the theory of material imagination: "Is it even necessary to point out that the adjective most closely associated with the noun *air* is *free*?" Through images from aerial poets, from dreams of flight and of falling (the fall being "an inverted ascent"), Bachelard illustrates an "ascensional psychology," a psychology of rising flight through the air that liberates us. He speaks of therapeutic experiments in which patients are invited to think of themselves ascending, becoming weightless. That is why "imaginary" air in particular is the "hormone" that "allows us to *grow* psychically." As we breathe more freely and become lighter, we imagine new realities.

For Burton, on the other hand, it is real air that enlivens our imaginations, and it operates through a real hormone, or humour. Air, being invisible except for its effects (wind blowing through trees) and sometimes its contents (water vapor in the form of clouds or fog), seems to be the most subtle of the four classical elements. The humours more or less extend this four-part typology to the functions of the body. The theory of the humours is a meteorology of the self, but it is not distinct from the meteorology of the outward weather; air, after all, circulates through us. Hence, one of the most powerful ideas embedded in Burton's work is that of equivalence, between our internal "temperature" and the external, and between our corporeal and spiritual selves and the natural world. To ignore these interdependencies and effects is madness—melancholy—just as it would be crazy not to be mindful of the clouds of devils you might be inhaling at any moment.

What is a humour? "A humour is a liquid or fluent part of the body, comprehended in it, for the preservation of it; and is either innate or born with us, or adventitious or acquisite." There are "four first primary humours": blood, phlegm, choler, and, last but not least, melancholy. Each has its own character, though all "be comprehended in the mass of blood," and they each "have their several affections." Blood, for example, is "a hot, sweet, temperate, red humour," whose function is "to nourish the whole body, to give it strength and color." Phlegm is cold and moist, and acts as a kind of lubricant of our working parts: "His office is to nourish and moisten the members of the body which, as the tongue, are moved, that they be not over-dry." Choler is "hot and dry, bitter," and "it helps the natural heat and senses, and serves to the expelling of excrements."

Which leaves melancholy. This humour is "cold and dry, thick, black, and sour, begotten of the more feculent part of nourishment, and purged from the spleen." It—to revert to the automotive analogy—acts as a kind of radiator fluid or coolant: "a bridle to the other two hot humours, blood and choler, preserving them in the blood, and nourishing the bones." In this typology of the humours, the emphasis is on balance and coherence; everything relates to everything else, and every element has a correspondent and countervailing opposite. "These four humours have some analogy with the four elements, and to the four ages of man." There are secondary humours, like "ros" and "gluten," and there is spirit as well: "Spirit is a most subtle vapour, which is expressed from the blood, and the instrument of the soul, to perform all his actions; a common tie or medium between the body and the soul, as some will have it." Each humour has a particular seat or organ in the body. All of these participate in the balancing of hot and cold, dry and wet, earth, air, water, and fire. Melancholy therefore isn't always bad, and the word has two meanings, the disease melancholy being in a sense "too much of a good thing"—an overabundance of the melancholy humour.

But this is exactly why melancholy has so much to do with air.

If air is taken by blood vessels from the lungs to the heart to moderate our spirits ("spirits are first begotten in the heart, which afterwards by the arteries are communicated to the other parts"), then it has a very similar function to that of melancholy, which moderates the hot humours. In fact, "bad air" is one of the causes of melancholy. Furthermore, because the brain is "a privy counsellor and chancellor to the heart," any upsetting of these fine balances among humours is likely to be felt on up the line, eventually unhinging our mental faculties.

The idea that Burton develops of the vast devastating power of the melancholic humour, toppling individual minds and bodies as well as kingdoms, might not be any more foolish than the current notion of stress, which no one has ever seen except in its effects and which is called on to explain any and all human maladies, including, in some cases, acts of madness.

Yet the theory of the bodily humours is not one you are likely to hear mentioned in your doctor's office, and it is doubtful that anyone now living conceives of the relationship between air and the human body quite the way Burton did. For example, since the melancholy humour is cold and dry, places with cold and dry climates increase its level in the body (through respiration) and cause their residents to be melancholy—(the meteorology of the environment meets the meteorology of the soul.) At the other extreme, in hot climates, one's melancholy is unduly dried up, which paradoxically leads to madness—melancholy: "Hot countries are most troubled by melancholy, and . . . there are therefore in Spain, Africa, and Asia Minor, great numbers of madmen." Denizens of hot climates also become quarrelsome, or rather choleric, Burton says; this must be because melancholy moderates the hot humours (blood and choler), and as a result of the heat, these unfortunates are always running low of humourous melancholy. Their hearts are insufficiently refrigerated because they are, literally, full of hot air.

Refrigeration of the heart? Madness owing to bad air? From a modern scientific standpoint—the birth of which, ironically, Burton assisted at—these ideas have no value because they lack scientific validity. What Bachelard is saying, some 350 years later, is that rationality is not the only scale of value. Modern science may yield technological advances and material improvements, but it is far from telling us how to correct other human ills—still less, how to live. Burton's words echo down the ages: "So much science, so little conscience." They also resonate with Bachelard's effort to resurrect a spiritualized materiality. He wants to free us from mere perception and the psychic torpor of despiritualized materialism. Through the material imagination, everyone can be a true poet and a true creator. Air is freedom, and a poem about air and a dream of flight, for example, is "essentially an *aspiration toward new images*." For Bachelard, new images are tangible things, not new pictures depicting something else. His use of the word *aspiration* is deliberate, showing how even on the level of language, our deepest desires move on air. Similarly, he points out that the Latin word *volo* means both "I want" and "I fly." Aspiration leads to inspiration, and we take flight. "The poet of fire, of water, or of earth does not convey the same inspiration as does the poet of air."

But today we live in a dangerous time. We are at risk of no longer taking flight. Technological culture suffers from a lack of imagination. (To invent is not the same as to imagine.) Bachelard, the philosopher of science, was alive to the criticism scientific colleagues would make of the "material imagination": "I will be told once again, that I have stopped being a philosopher to become a mere collector of literary images. But I will defend myself by repeating my thesis: the literary image has its own life; it *moves as an autonomous phenomenon above profound thought*."

But even the modern poets will be against Bachelard, because, "Alas! the poetic super-ego is influenced by literary criticism," and "Isn't it obvious that literary criticism has made an almost unconditional treaty with 'realism,' and that it takes umbrage at the slightest attempt at idealization?" The mulish tendency of cynical critique—critique for its own sake—will object to "rising," "ascension," and "transcendence." Artists are now seen as producers, the artwork an object in an economy of consumption. "Transcendence" is a dream of the age of faith, now shown to be dangerous—the past seems littered with dangerous dreams. We must stick to what is "real." Reasoning thus, the critic supports "a suppression of the ideal, a suppression that is thought to be based on a reality—but which is only a reality of repression, which is also considered to be founded on reason, which is only a systematizing of repression."

Bachelard implies that reason and its "truths" seem self-evident because they have systematically repressed all other possible scales of value. Nietzsche made the point in "On Truth and Lie in the Extra-Moral Sense" that rationalism is a system and "truths" its by-products. What ideas or perceptions the system refuses to accept as input obviously cannot lead to "truths," but that does not mean they cease to exist. A defense of the theory of the humours and of the material imagination as airborne modes of ascension, understanding, and psychic enrichment will seem scandalous only as long as one accepts this suppression of the ideal in favor of reason. At the time Burton was writing, reason had not so firmly established itself that the imaginative mode of grasping reality had been relinquished. When Bachelard came to write of the material imagination, reason had so routed the opposition that a vacuum had been created, an imaginative and spiritual impoverishment, preparing the way for the return of the repressed—the fresh air of the ideal and the suppressed aspiration for new images.

Bachelard's theory of the elements closely shadows Burton's theory of the humours. The poet of fire will be passionate (choleric); the poet of water will be meditative (phlegm); the poet of earth will delight in the physical (blood); the poet of air will be melancholy. It is in their thinking about melancholy that these two renegade philosophers meet. The theory of the humours is bound to suffer the same fate as "material imagination" at the hands of technological rationalism, which will insist that the humours have no biological or epidemiological value. They are somatically false; they are not "real." This is because of the technologization of the body itself, which is seen as a machine, a collection of parts. We don't imagine ourselves anymore; doctoring has nothing to do with imagining. With cloning, the body even becomes a manufactured object.

Evaporating and condensing like the ambient air, we are air. We live our humours. But unfortunately our humours are out of accord, and we have become melancholy, mad. Unable to bear that "ambivalence," that equilibrium between dream and reality in which the truth is not known, in which reality is less understood than sensed, we settle for being "mute slaves of an inert matter" and the materialism that is its theory. We (culturally, institutionally, personally) could use a breath of fresh air. The imaginative possibilities of the aerial dimension are there for those who, ready to bear the melancholy of the heights, begin once again the ascensional journey.

Song of the Andoumboulou: 50

Nathaniel Mackey

Fray was the name where we came
   to next. Might've been a place,
might not've been a place but
     we were there, came to it

                    sooner
   than we could see . . . Come to
so soon, it was a name we stuck
   pins in hoping we'd stay. Stray
was all we ended up with. Spar
    was another name we heard

                 it
    went by . . . Rasp we also heard it

                    was
   called . . .   Came to it sooner
than we could see but soon enough
    saw we were there. Some who'd
come before us called it Bray . . .

Sound's own principality it was, a
   pocket of air flexed mouthlike,
meaning's mime and regret, a squib of
    something said, so intent it
seemed. At our backs a blown

                  conch,
   bamboo flute, tropic remnant,

                Lone
   Coast reconnoiter come up empty
but for that,   a first, forgotten
   warble trafficked in again even so,

                  the
    mango seed's reminder sent to what
   end we'd eventually see . . .

             We had
   come thru there before we were

told. Others claiming to be us had
  come thru . . . The ubiquitous two lay
bound by cloth come down from on
              high,
hoping it so, twist of their raiment
            steep
  integument, emollient feel for what
  might not've been there. Head in the
    clouds he'd have said of himself,
             she'd
  have said elsewhere, his to be above and
    below, not know or say, hers to be
    alibi, elegy otherwise known . . .

Above and below, limbo what fabric
intervened. Limbo the bending they
    moved in between. Limbo the book of
            the
    bent knee . . . Antiphonal thread
    attended by thread. Keening string
      by thrum, inwardness, netherness . . .
            Violin
      strings tied their hair high, limbo
    the headrags they wore . . . The admission
    of cloth that it was cover, what
    was immanent out of reach,   given
            what
went for real, unreal,
        split,
  silhouetted
redress

Over the next hill we ran into
rain, were suddenly wet. The
    two who'd fallen in with us threw
        back their heads, drank it
        in. "No more shadow mouth,"
                                they
        shouted, which was true, the
        mango seed was way behind us
            now. No tropic sweetness's mock
            proffer, it was north we truly
were, circling counterclockwise,
                            in the
        clear, close if not quite, lifted as
we circled it seemed. A conjoint
    lowing emanated from the ground,
chthonic voices mounted higher,
                        skyward,
        Spite Choir, the chorale we'd heard
        so much about of late we took it
to be . . . Hallelujah Hill it was we
        were now the near side of, an
        allegorical hill we'd heard we'd
                            come
        to, climb, come away from forgetful,
            rummage whose underside. Some we
        met said it was only a trap, rapt-
            anagrammatic diminution we were
        shadowed by, mango seed retreat
notwithstanding, demiurgic trick . . .
                                We
        paid them no mind, plowed ahead, un-
        impressed. The two who'd fallen in with
        us led the way. It wasn't wandering what
we did, we circled, an earthbound orbit
                            wanting
        out we went up on, low Saturnian shout,
                            rings

we walked. The two who'd fallen in with
us led the way, mapped our way, no sooner set

                                      our

course than they
were gone

From somewhere near the well we
heard singing, voice an unheard-of
porridge, capsicum and roux,
       "Shadow
mouth," it lamented, "shook my tree."
 More than could be carried we caught,
were whisked away by, movement the
one mooring we knew. Eyes tightly
shut, one tear squeezed out by the
  impending end, what we wanted
          was to
  endlessly verge on exit, angling
  out, tangent to circle's edge, on
  our way where we'd be the last to
  say . . . Let all edges converge, it
         seemed
we said, cut away would-be end. Shell's
 edge, knife's edge, pearl we'd be
 prompted by, refugees from where
 likewise last to say . . . It wasn't
wander what we did, we circled. Frayed
        at its
 edge though it was, wheel of soul, verge
 we were driven by . . . Verge that we wanted
 verge was the song we sang had there been a
song we sang. No song left our lips.

 Nonsonant, we rounded circle's edge,
nonsonant ring shout, verge our muse
 and mount. Verge that we wanted verge
we bordered on singing. No song left
 our lips. It wasn't sing what we did,
       we
 circled. Song was the porridge voice's
privilege. "Shadow mouth," it repeated,
        "shook
my tree . . . " Sparks rose near the well, an

extinguished fire, hung like a signal
  or a sign of moving on, a symbol, some
said, showing forth . . . "Post-ecstatic"

                       was

a word we heard, "copacetic" a word we
  heard, "After ecstasy what?" a question
posed in smoke . . . It wasn't smolder what we
  did, we burned wanting verge,   verge

                            riding

  our legs, we
bore thru

Verge that we wanted verge kept
insinuating,   song we'd have sung
    had there been one,   anthem
circling assumed. It was a healing

                                        song
    we sang had there been a song we
    sang, swirling water we intimated
        wet our feet . . . Momentarily two, I
    levitated, hoisted by thread Ananse-web

                                        thin . . .
    Wholly elbows and shoulders I was, it was
Nudge I knew I was in . . .   Pulled in
        early, a taste in my mouth I wanted
    out, threw back drinks hoping to
        wash it away. B'Hest tended bar,
tilted my head back again and again.

                                        It
        wasn't the buzz but the feel of wash

                                        I
wanted. Water would've been enough . . .
    Elbows and shoulders got forearms,
    hands, lifted drinks. Water by itself
        would've done . . . Powder coated
        my tongue, I spat cotton. An
    astringent puckered my cheeks,

                            an
        aftertaste . . . Limped, went limp,

                                    down
        and up at the same time,

                            first
    and forever bone-to-be-gutted,

                            break-
    intimated
flight

Tore the earth and tore the air
we heard later, we the instructive
   two they went on about so. Tore
their throats wanting to swallow
   the sky, tore their clothing,
                              word
reached us farther along the way
   they'd been brought back to earth,
                           now
      no longer sure they'd ever left . . .
      Tore their hair, wrung their
      hands, felt Ogun in their grip,
                          tore
         skin. It wasn't limbo they were
      in albeit limber their legs and
      arms got, loose the realm they'd
   move thru next . . . Tore antiphonal rope,
                              hit-
      upside-the-head welter, gangly the walk
      they now walked. Legs made of sea lurch of
      late, tore the sea, we the weight on their
                              backs
      unbeknown
   to us

Danghi Korwa, *Untitled* (37)

The Mystery of the Hills

Franck André Jamme

Sometimes we go very far in search of things.

This is a kind of story. It comes from the depths of India, Bhopal, the capital of Madhya Pradesh, and, more specifically, the Bharat Bhavan, an enormous "cultural complex" on the edge of the Upper Lake—a center that holds, amongst other things, one of the richest and most beautiful collections of Indian tribal art. It is very little known by Westerners that India is a land of tribes—really you aren't expecting it, when you think, too rapidly, of the sophistication and the baroque of the subcontinent (or in fact of the huge buildings of that other great city in the south, where thousands of computers deal at a distance with the accountancy of a few big Western banks). There are perhaps five hundred tribes, more or less lost, more or less alive, more or less free from contact with the "modernity" of the huge country. But still, today, five hundred. Each with its culture, its traditions, its arts.

So in Bharat Bhavan in Bhopal, about three years ago, the rooms were, as so they often are in Indian museums, let's say, in the process of reorganization: there reigned, in fact, a pleasant disorder. And there, in a corner, after a series of quite disquieting local masks and an imposing village totem, which probably represented a tiger, but a tiger riddled by so many monsoons that in fact it no longer looked like a tiger, I was suddenly in front of three pieces, slightly dusty, but carefully framed, one of them on the wall, the two others just propped against the wall at the foot of the first one, as though they were looking after it. Three very strange things.

They were not paintings, nor really drawings, although from time to time I could find in them, here and there, a figure, an animal, or an object. It was writing rather, large manuscript pages imitating, you would have thought, an Indian script—because, at points, it often looked like Hindi, with the frequent horizontal bar above but without any precisely recognizable letter from that language. Particularly striking was the general way in which these signs had been traced, the astonishing appropriation of the space that I had before my eyes. With the lines that frantically ran one over the other, one under the other, or sometimes simply got lost in archipelagoes. Not forgetting the wide spaces that let the movements breathe, and the constantly agitated columns or bunches of signs. Fascinating.

I asked of the first warden I came across whether I could take photographs. No way. I went back to the main entry where I had noticed in passing a tiny shop selling postcards and books and asked. No they had no postcards of the exhibit, but there was a small catalogue, and they would go and get it quickly. A good hour later, another warden came back triumphantly with a packet under his arm. I bought a copy of the precious volume and went back to the hotel.

On the way in the "scooter," I wasn't able to read the catalogue, but I flipped through it, saying to myself that *The Magical Script* (that was the title) revealed yet another traditional art, repeated from generation to generation, as so many are in the tribes. I let my thoughts flow. Perhaps it was a series of messages, who knows? Perhaps in this culture, when a man was in love, he would send, by a friend or a cousin of the person who so pleased him, one of these expansive pages of writing. Love letters: I sat dreaming. At the next crossroads, I envisaged, rather, missives to the gods. With vows, with prayers—and even a few reproaches. And carried on dreaming.

As soon as I got to my hotel room, I devoured the little book. I learned a lot. First of all, that it was not a traditional activity. Something quite different, another story, as we shall see, which might begin like this. . . .

There was in India a remarkable man who is now dead. He was called Swaminathan. He was first of all a painter and, impassioned as he was by the tribal cultures of his country, he was also a patron of that same Bharat Bhavan of Bhopal. He liked going to meet the tribes and particularly those of the state in which he lived. One day, in 1983, after a lecture by an anthropologist called Kalia, who had a wide knowledge of the region, he decided to make an expedition toward the northeastern horn of Madhya Pradesh, in the Raigarh District, amongst the Korwa Hills.

When he arrived he found the Korwa touching and enjoyed their laughter. And certainly he appreciated their taste for *mahua,* a wine made out of flowers that they consume in abundance. On the big dais of the village, the interpreter did his best;

nevertheless Swaminathan wanted to say more to those men and women he had before him: he took out a little notebook from his waistcoat pocket and started to make sketches of the faces surrounding him. One of the villagers then asked him for a bigger sheet. Others came, laughing, and made it understood that they wanted to be in on the game. From the start, it seemed to amuse them a lot. There was paper, paint, brushes, and markers in the jeeps, and they were fetched. Sheets of paper were stretched out on the wood of the dais with stones on all four corners, and a good dozen of *adivasis,* men and women, set to work.

All the members of the expedition expected to see pictures of animals, and a few human figures, surge from the villagers' hands—that is what the populations of the region usually draw. But in this instance the unlikely occurred. Leaning over their pages, they started immediately to trace the kind of signs I had seen at the museum. Forms of writing that were different from one Korwa to the next, nothing but lines or clusters of signs: forms of writing. Just discreetly sprinkled from time to time, rarely, with animals, human figures, and a few objects, bows or arrows. This went on for hours. And what is more started up again the next day. And so, every day, for a whole week, during which there was a regular resupply of paper for the villagers—when there wasn't enough there had to be a trip to the nearest township to get more. When it had all finished, the full sheets were gathered up—with the empty bottles of *mahua.*

Different forms of writing. Cautious approaches to Hindi, lines littered with bows and arrows, huge open goatskin bottles with unheard of letters, pure mescalinian repetitions, frantic scribblings. So, varied. Each to his or her own world.

And if I say that what was produced was incredible, it is because the Korwa are illiterate and never see writing outside Sanna or other large villages where they go to sell vegetables and wood, or when they receive the visit of a district bureaucrat, who inevitably takes notes or brings out forms.

For me the trip was almost over. I had to get back to France. With, at the back of my eye, as though they were tattooed on it, the three pieces from the museum and the reproductions from the little catalogue. Passing through Delhi, I tried to get more information. Once in Paris, I sent off letters, more than one of which were like bottles cast into the sea. I had found the names of some of the artists and their villages. I even wrote over there, to the hills. The letters were later found at the bottom of huts in Laldarpat, in Pendrapat: they were never read. And then one day the ball finally bounced back—from the other end of the world and back. One of the new directors of the Bharat Bhavan, Akhilesh, replied to my question: yes, he was game for an expedition to the Korwa.

Which had to be the proof of a deep interest in these people and, since we didn't know each other at all—a no lesser proof of confidence. For the journey from Bhopal up to the North of the Raigarh District, on the frontier of the Bihar, is really an adventure. A thousand kilometers (which counts for twice the distance in Europe) by train, then by jeep, and finally on foot. Not to mention the inherent difficulty of all the restricted areas, among them, Raigarh: Let's say an often tense socio-political (or rather socio-ethnic) situation.

But what is known of this small population? The Hill Korwa are few in number, about five thousand. They speak *sadari,* a dialect of the family of languages known as *mundari.* According to their foundation myth, the first Korwa was a daruna, a scarecrow. As the legend goes, a god and a goddess were wandering in the countryside when they noticed in a field a scarecrow holding in one hand a bow, and in the other an arrow. The goddess then asked her companion on the walk to give life to this being made of rags and twigs, which the god immediately did.

For a long time, the forest-dwelling Korwa were solely hunters. Using bows and arrows, they hunted birds, deer, stags, wild boars and bears. But with the chopping down of trees, the bowmen gradually had to convert to agriculture. They were given carts and buffaloes and asked to become farmers of vegetables and maize. The large-scale destruction of the trees and the further harm caused by farming had a pernicious effect. The region is home to herds of wild elephants, and nothing attracts these pachyderms more than the sweet smell of maize. The giant beasts, no longer stopped by the barrier of woods, come regularly to ravage the plantations—as well as the huts and their contents, trampling everything under their path. Each year a few villagers are crushed to death, particularly children.

Once a week the little men go down into the valley townships (particularly Sanna, which is the biggest market in the region) to sell wood and vegetables and to buy what they are in need of, amongst other things, salt and *mahua.* Sell or barter. The custom is, for a liter of *mahua,* twenty-five or thirty kilos of wood.

Their gods have names which sound like song: Khuria Rani, Dulja De, Thakur De. Their rituals involve sacrifices—chickens, sometimes goats. In spite of the presence of Christian missionaries in the Raigarh District and the insistence of Hindu proselytism, they have always proved rebellious to conversion and have hardly been "contaminated" by either religion. Each village has its *baiga,* or sorcerer, who is also the community doctor, and this has often had ill effects. In particular, the *baigas* don't try to stop pregnant women from puffing on their *biddhis,* or from drinking. In face of the hecatomb (before, during, and after giving birth),

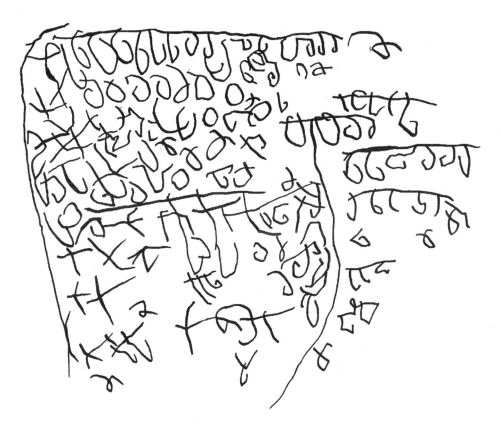

Dhingra Korwa, *Untitled*

the authorities decided not to apply to the Korwa the national rule over birth control. And what more? They bury their dead, which is unusual in India. Most often in the patio of their huts. They don't use the space of the "tomb," they don't put anything there, they don't stand there too long. Lots of blanks, so lots of empty spaces in the courtyards.

They get up early. At dawn, they go off to work in the fields. Except for the old women, everyone follows, including the kids. In fact, they never stop moving, running around.

Up, down, up, down. On the scenic railway of the hills, a healthy adult can cover forty kilometers a day without any problems. Great walkers, who also like walking fast. They even have a word for it, letting out a cry: "Dalongo! Dalongo!" And off they rush. And then they drink. A lot. Almost all the time. They love it. Above all *mahua,* but also *bandi,* a rice beer. You quite often come across a Korwa stretched out right in the middle of a path, or under a tree, just a bit dreamy, just a bit not with it. They also laugh. Almost all the time. These things go together.

And, I'll say it again, illiterate.

Besides, the country is beautiful. Even very beautiful. Nothing boring, nothing flat. Hills, an intense green. Fields of red earth. From time to time a hut, ochre and straw. The golden yellow of the maize. And everywhere the smell of flowers.

Just a word about *mahua.* Quite unique that wine. Quite strong, almost harsh, transparent, you might think it was water. Quite strong, but it slips down the throat very well, too well, too quickly. You don't forget once you've known it, and it gets a grip on you: stay a few weeks and you are hooked to the forest. I remember Udayan Vajpeyi recalling his experience in the hills in 1993. He had gone there to help make a film: "Three or four glasses and everything becomes clear. The sky seems to be washed clean, the air becomes translucent, the mountain greener than green, as though everything had been repainted in its own colours. As though the world were suddenly made of the soft moist earth of dreams."

The Korwa don't know how to read. Or write. You keep coming back to this with them. Given that they write nevertheless. In spite of all. That they write instantaneously. Even if the signs they make are not real letters, even if no one will ever be able to translate those lines into another language. Ever.

There is some sort of mystery in the hills, that's for sure.

When you ask them why they make these traces, they are evasive in their replies, rather enigmatic. They don't really know. It seems that they don't really like whys. They say, "That's the mark of the Korwa," and burst out laughing, then mumble, "It makes you feel good." They take the first *mandal* to hand and play on it. They admit, occasionally, a little more. "It's to tell of our suffering," just before starting to laugh again. Joy and sorrow all in the same bag. On the back. In the head. Muffler for poverty, muffler for sorrow, and sometimes, joy.

And then why, after all, a why?

Certainly Swaminathan was not wrong when at first he scratched his head, recognizing that "it will require a lot of investigation and inquiry by anthropologists as well as psychologists to arrive at some plausible explanation and conclusion." Again, when he spoke, cautiously, of an "invocation." But then, an invocation to writing itself. I'll go with it.

It could be a call. With the simple means of graphic signs which are blind, without rules, lacking in any clear sense, as though prenatal, the Korwa seem to be asking just for writing to appear, to come.

A variant. Perhaps they have so much desire for writing they end up tracing it anyway, without knowledge, as it comes. On the paper there would in this case be of course no more than a dream. Too bad if it can't be communicated: at least it would be there, in black and white (or in another color).

Or perhaps it's a kind of trance. Instead of making cries, gesticulating and dancing like any good shaman, the Korwa simply write. Men and women sit in front of their sheets of paper, start to emit strange sounds, wriggle and wiggle, let fall a word or two, roll their eyes and twist their necks, as though they were suddenly beginning to do something else—or rather as though they were living what they were "writing." Shamans without coded rituals legible by all. Shamans without shamanism. Untamed shamans.

Certainly Archana (the only person to have known both expeditions, those of 1983 and 1996) is right in suggesting that these *adivasis* possess so little, are content with so little, that they have made a lot of what they have—almost everything. Which is why, when they find themselves in front of the white page, a thought unfolds its petals on the surface of their minds: the only thing they are lacking, "the only palpable absence," as Archana says, is written language. I agree with that too.

So, I came back. There I am in France, on the steps of a worn stone staircase that leads down into a huge, rather derelict barn. So far from the inspired hills. I put down on the floor the pieces that will soon be shown in New York. So far from the inspired hills. There is not the slightest noise—just, outside, in the garden, a few birds having a serious discussion. I let my gaze slip from one huge page to another, and then I collect them all together in the depth of my eyes, which I close. I no longer see anything. Only a bundle of qualities in a quiver.

To start with, it's clear these people have talent. Their ways of taking hold of space, above all, I have already taken note of, those great breaths of the void, those rhythms so open and so supple, of signs and lines, make you think that they are professional artists. This is so true that Swaminathan, at the moment of his discovery, did not believe his eyes (and yet amongst the tribes he had seen a lot of things). When I returned from the Madhya Pradesh, the first people to whom I showed these writings sometimes struggled with the fact that these enormously skillful dances on paper had been traced by aborigines who hardly ever touched a pencil. (*On gift.*)

Strange too, how abstraction comes naturally to them. Instead of a story recounted (with its places, its personnages, its paths, its animals, or its trees, as you find in most of the Indian tribes who draw), with the Korwa, you are in contact with the transcription of pure and simple movements of the mind, almost as though they were captured at their birth. (*On natural abstraction.*)

Also striking is how it all seems to go so fast. Alive, boldly drawn are these signs. Intuitive, spontaneous. Quick as arrows. Like the arrows in their quivers, now hung in their huts. Haste would be their mark. They drink their glasses at a throw. They like walking with big steps: Dalongo! Dalongo! And when they have to choose between pencils, brushes, pastels, markers, they usually seize the last "because it's faster," said one of them. (*On love of speed.*)

Again, arrows. "Broken arrows," as Archana says. And even goes much further. "When they write I think they take their pencil for a bow. They don't write in fact: they shoot. They shoot signs which are arrows. That must be it." Quite right.

Blind arrows. Huge writing blindfolded. Letters with no fixed face. Which finally merely give us a taste of, a love of the plasticity of their rolling and pitching.

Incredible too, how these pages seem reflexive. Nothing to construct, nothing to think upon. You take just what moves in your head when what moves appears. At the root. And you put it back together as it is. It might be fun to think (affectionately) of André Breton faced with these pieces, a lively Breton, as usual on a relatively high horse, head barely thrown back, plume thrown back, running in search of tribal automatism, so strongly is he overcome by the "subitism" of this Korwa writing. But the spontaneity is that of signs that never turn back on themselves, are never corrected, of those lines that are truly automatic. (*On unintentional automatism.*)

In the end, there would be no constricting bonds. Because the Korwa, at the end of the day, just don't care, because they sometimes whip up the whirlwind of signs to the point where it becomes no more than a single liberated line, changed

Lahangi Korwa, *Untitled* (34)

only by banking, turning, coming back on itself, managing to devour the world of the page with marvelous scribbling. *(On permissiveness.)*

Do we know things which are not necessarily similar, but at least close, elsewhere, somewhere? Of course there would be all those Western painters, of this century, who also "wrote" untranslatable sentences: Klee, Degottex, Twombly, Michaux and others. There are also other figures such as Pierrette Bloch, for example, but here we already talking about an artist, or Gaidiliu, a woman active in the struggle for the Naga, who filled her notebooks with a writing "which does not match any known language." Or the mountain peasant woman—Helen Smith from Switzerland—who calligraphed a language that she herself called "Martian." Things so close? But I am not sure.

In any case, for Michaux, Twombly, and the others I don't put much credit by it, because as time goes by, these parallels (and inevitably, these comparisons) seem more and more hollow. Finally, I get fed up. During the many years in Europe and the United States that non-Western works have been shown alongside those of contemporary artists from rich countries, it is obvious that it has been accepted that the former are finally visible and exist, and that the universality of the human artistic gesture is from now on evident, but at the same time people have continued to judge everything on the single yardstick of the north. Whereas we ought to come to grips with these arts from distant parts on their own terms. So long as we don't go into some fatal telescoping—that of course always exists—you could almost start worrying about "coupled" linkings. Alone the Korwa and the others. Just them on the wall and us looking at them. Just one and the other. Just two.

And you might rather dream of being naked. In spite of that invisible baggage of culture, of personal and collective history that we all cart around on our back, which carts us around and weighs us down. Dream of being naked, arriving in front of any work as little deadened as possible by the subliminal fog that comes not only from the taste of the day but also from knowledge, from the past.

Naked as the mind of these men and women who at first can be frightened if they spot you on the turn of a path, and run off to hide in the forest. But who, a moment later, if you look nice, come out of the woods, take you into the courtyard of their hut, offer you *mahua,* down their glass in a stroke, look you in the eye, and laugh.

Look you in the eye
See you
See the world within you and around you

See themselves in the mirror of your pupils
Finally laugh not only at you but at the world and themselves
All in one.

I said this at the beginning: sometimes we go very far in search of things. In fact we have known for a long time that there is nothing more human than laughter. Nothing is more purely human than the laughter of an inspired person, any inspired person—so too those inspired people from the depths of India.

Inspired by what? Finally, I don't know. That is the mystery of the hills.
And of the world.

E. M. Bidwell, from *Harper's Weekly*, Jan. 17, 1887

From *Of Walking in Ice*

Werner Herzog

In 1974, at the age of thirty-two, Werner Herzog, one of the most renowned German directors of the twentieth century, set out on a walk across Europe to visit his friend Lotte Eisner, a well respected film critic, who he believed to be dying. Traveling with little more than a coat, he spent the majority of the trip thinking, breathing, and looking up at the clouds. Throughout the memoir, Herzog is rooted to the ground and at the mercy of the sky. As he walks, his thoughts take off in the breeze.

At the end of November 1974, a friend from Paris called and told me that Lotte Eisner was seriously ill and would probably die. I said that this must not be, not at this time, German cinema could not do without her now, we would not permit her death. I took a jacket, a compass and a duffel bag with the necessities. My boots were so solid and new that I had confidence in them. I set off on the most direct route to Paris, in full faith, believing that she would stay alive if I came on foot. Besides, I wanted to be alone with myself.

Sunday 24 November 1974
Fog outside, so icy cold that I can't describe it. On the pond swims a membrane of ice. The birds wake up, noises. On the landing my steps sound so hollow. I dried my face in the cottage with a towel that was hanging there; it reeked so bitterly of sweat that I'll carry the stench around with me all day long. Preliminary problems with my boots, still so new that they pinch. I tried using some foam and, with every movement wary like an animal, I think I possess the thoughts of

animals as well. Inside, beside the door hangs a chime, a set of three small goat-bells with a tongue in the middle and a tassel for pulling. Two nut bars to eat; perhaps I'll reach the Lech today. A host of crows accompanies me through the fog. A farmer is transporting manure on a Sunday. Cawing in the fog. The tractor tracks are deeply embedded. In the middle of a courtyard there was a flattened, gigantic mountain of wet, filthy sugar beets. Angerhof: I've lost my way. Sunday bells from several villages in the fog all at once, most likely the start of church services. Still the crows. 9 o'clock.

Mythical hills in the mist, built from sugar beets, lining the path through the field. A hoarse dog. I think of Sachrang, as I cut off a piece from the beet and eat it. It seems to me the syrup had a lot of foam on top; the taste brings back the memory. Holzhausen: the road emerges. By the first farm, something harvested, covered with a plastic tarp anchored with old tyres. You pass a lot of discarded rubbish as you walk.

A brief rest near Schoengeising, along the River Amper; matted countryside, meadows at the edge of a forest, rimmed with rifle ranges. From one rifle range you can see Schoengeising; the fog clears, jays appear. In the house last night I peed into an old rubber boot. A hunter, with a second hunter near by, asked me what I was looking for up there. I said I liked his dog better than him.

Wildenroth, 'The Old Innkeeper' tavern. Followed the Amper; empty, wintry weekend cottages. An elderly man, enveloped in smoke, was standing by a dwarf pine, filling the house of a titmouse with food; the smoke was rising from his chimney. I greeted him and hesitated before asking if he had some hot coffee on the stove. At the entrance to the village I saw an old woman, small, bow-legged, madness etched across her face; she pushed a bicycle, delivering the Sunday paper. She stalked the houses as if they were The Enemy. A child wants to play pick-up sticks with plastic straws. The waitress herself is eating right now; here she comes, chewing.

A harness hangs in my corner and within it a red streetlight is mounted for lighting, a loudspeaker above. Zither music and yodelled "Hollereidi," my beautiful Tyrol comes from on high.

A cold mist rises from the ploughed, broken fields. Two Africans were walking in front of me making thoroughly African hand gestures, deeply engrossed in conversation. To the very end they didn't notice I was behind them. The most desolate thing was the palisades of Hot Gun Western City, here in the middle of the forest, all dreary, cold, void. A railway that will never run again. The journey is getting long.

For miles across open fields I followed two teenage village beauties along a country road. They were walking a little slower than I, one of them in a miniskirt with a handbag, and after several miles I steadily drew close to them. They saw me from far away, turned around, quickened their pace, then stepped a bit slower again. Only when we were within reach of the village did they feel safe. When I overtook them I had the feeling they were disappointed. Then a farmhouse at the edge of town. From a distance I could see an old woman on all fours; she wanted to get up but couldn't. I thought at first she was doing something like push -ups, but she was so rigid she couldn't get up. On all fours she worked her way to the corner of the house, behind which were people who belonged to her. Hausen, near Geltendorf.

From a hillock I gaze across the countryside, which stretches like a grassy hollow. In my direction, Walteshausen; a short way to the right, a flock of sheep; I hear the shepherd but I can't see him. The land is bleak and frozen. A man, ever so far away, crosses the fields. Phillipp wrote words in the sand in front of me: Ocean, Clouds, Sun, then a word he invented. Never did he speak a single word to anyone. In Pestenacker, people seem unreal to me. And now the question, where to sleep?

Monday 25 November
Night near Beuerbach in a barn, downstairs serving as a shelter for the cows, the earth like clay and deeply trampled down. Up above it's passable, only light is missing. The night seemed long, but it was warm enough. Deep clouds outside sweep by, it is stormy, everything seems grey. The tractors have their headlights on, although there is light enough. After precisely one hundred metres a roadside shrine with little pews. What a sunrise behind me. The clouds had split open a crack; yes, a sun like that rises bloodied on the day of Battle. Meagre, leafless poplars, a raven flying although missing a quarter of his wing, which means rain. Beautiful, dry grass around me, rocking itself in the storm. Right in front of the pew a tractor track in the dirt. The village is dead silent, telling of deeds done from which it refuses to wake. The first traces of blisters on both heels, especially the right one; to put on my shoes requires a great deal of care. I must get as far as Schwabmuenchen, for plasters and money. The clouds are drifting against me. My God, how heavy the earth is from the rain. Turkeys are screaming with alarm from a farmhouse behind me.

Outside of Klosterlechfeld. Now I can see that the Lech would be no problem, even without a bridge. The terrain reminds me of Canada. Barracks, soldiers in huts of corrugated iron, bunkers from the Second World War. A pheasant took off just three feet in front of me. There's fire from a tin oil drum. A deserted bus stop; children have emblazoned it with coloured chalk. Part of a wall made of corrugated plastic is banging in the wind. A sticker here notifies us that the power will be turned off tomorrow, but for a hundred metres around me there's nothing electrical to be seen. Rain. Tractors. The cars still have their lights on.

Raging storm and raging rain from the River Lech to Schwabmuenchen; nothing noticed but this. In the butcher's shop I stood waiting for ages, brooding Murder. The waitress in the inn understood it all in a glance, which did me good, so now I'm feeling better. Outside, a radio patrol car and police; I'll make a long detour around them later. While changing my large note in the bank, I had a definite feeling the teller would sound the alarm at any moment, and I know I would have run instantly. All morning I was ravenously hungry for milk. From now on I have no map. My most pressing needs: sticking-plaster, a torch.

When I looked out of the window, a raven was sitting with his head bowed in the rain, and didn't move. Much later he was still sitting there, motionless and freezing and lonely and still wrapped in his raven's thoughts. A brotherly feeling flashed through me and loneliness filled my breast.

Hail and storm, almost knocking me off my feet with the first gust. Blackness crept forth from the forest and at once I thought, this won't end well. Now the stuff's turning into snow. On the wet road I can see my reflection below me. For the past hour continual vomiting, only little mouthfuls, from drinking the milk too fast. The cows here break into a gallop quite unexpectedly. Refuge in a bus stop of rough stained wood, open to the west so that the snow blows into the most distant corner, where I am. Along with the storm and snow and rain, leaves are swirling as well, sticking to me and covering me completely. Away from here, onward.

A brief rest in a stretch of woodland. I can look into the valley, as I take the short cut over wet, slushy meadows; the road here makes a wide loop. What a snowstorm; now everything's calm again, I'm slowly drying out. Mickenhausen ahead of me, wherever the hell that is. Raindrops are still falling from the fir trees to the needle-covered ground. My thighs are steaming like a horse. Hill country, lots of woodland now, everything seems so foreign to me. The villages feign death as I approach.

Just before Mickenhausen (Muenster?), turning further west, following my instincts. Blisters on the balls of my toes give me trouble; had no idea that walking could hurt so much. High upon a telephone pole a repairman was hanging by his straps undaunted, shamelessly staring down at me, the Man of Sorrows, entrusting his weight to the taut strap while smoking a pipe. His glance followed me for a long time as I crept past below. Suddenly I stood rooted by my feet, then turned on my heel and stared back. All at once a cave in the craggy slope behind me howled down to the sea with its mouth wide open. The rivers streamed converging to their end in the sea, with the Grotesque also cowering in a crowd on the coast, just like everywhere else on this Earth. Overwhelming all was a sudden, strange, otherworldly whistling and whining in the air, from the gliders circling over the slopes. Further on, towards the rising sun where the thunder of faraway guns was rumbling, a radar on a mountaintop mysterious and forever taciturn, like a huge eavesdropping ear, yet also emitting shrieks that no one can hear, reaching into fathomless space. Nobody knows who built the station, who runs it, to whom it addresses itself. Or does the repairman strapped to the pole have something to do with it? Why is he staring after me like that? The radar station is often shrouded in clouds, then they scatter and the sun goes down, days passing as I stand there, and still the station glares fixedly at the ultimate edges of the universe. Over the mountain forest outside of Sachrang, in the last days of the War, an aeroplane dropped a metal device that was visible in the treetops by its flag. We children were certain the flag was wandering from tree to tree, that the mysterious device was moving forward. During the night some men went off and, when they returned at daybreak, they refused to divulge information concerning what they'd found.

Beautiful hilly countryside, a great deal of forest, all is still. A hawk screeches. On the prayer cross behind me is written: Ere night falls, all can swiftly change / And have a different face from early morn. / On earth a restless stranger was I born / In mortal danger, though in the midst of life. / Through Christ's blood for me my God pray send / Some good end to all this strife. / Our time is high, Eternity draws nigh. I noticed that the road turned more and more southward; thus, cross-country: Kirchheim, then, around a forest, dusk approaching. Obergessertshausen: no break, since it is almost totally dark. I'm rambling more than walking. Both legs hurt so much that I can barely put one in front of the other. How much is one million steps? Uphill towards Haselbach, in the darkness I can discern something but, stumbling forward, it turns out to be just a thoroughly filthy shelter for cows. The ground is trampled knee-deep in wet clay by their hooves, and my feet collect pounds of heavy, sticky clods of earth at once. On a rise before Haselbach two holiday homes, the prettier of which I break into without causing any damage. Remnants of a feast inside, it can't be long ago. A pack of cards, an empty beer mug, the calendar showing November. Outside there's a storm and inside there are mice. How cold it is!

Friday 29 November

Not a good night, therefore somewhat plaintive in the morning. Telephoned from the post office. An ugly, much-frequented road to Neufra over a range of hills. A direct route cross-country is hardly possible. A terrible storm up in Bitz, everything's covered with snow. Beyond Bitz, up a forested slope, a furious flurry of snow breaks out in the forest itself, the flakes circling down from above like a whirlwind. I don't dare venture into the open fields any more since the snow there blows horizontally. For many years they haven't had anything remotely like this, and it's not even December yet. A truck on a nearby road picks me up; it is able to advance cautiously, only at a man's pace. Together we push a stranded car out of the snow. In Trudelfingen I become aware that there's no way to proceed, the snowstorm's becoming a Madness. Tailfingen, again at an inn, I hang up my clothes. A standstill all day long, no movement, no thoughts, I've come to a standstill. The town is awful, quite a lot of industry, cheerless Turks, just one telephone booth. Very pronounced loneliness, also. The little one must be lying in bed by now, clinging to the edge of his blanket. Today, I'm told, they're already showing the film at the Leopold; I don't believe in justice.

Wednesday 4 December

An immaculately clear, cool morning. Everything is hazy on the plain, but one can hear life down there. The mountains, full and distinct in front of me, some elevated fog and, in between, a cool daytime moon, only half visible, opposite the sun. I walk straight between sun and moon. How exhilarating. Vineyards, sparrows, everything's so fresh. The night was pretty bad, no sleep from three o'clock on; in the morning, making up for it, the boots have lost their painful places and the legs are in order. The cool smoke of a factory rises calmly and vertically. Do I hear ravens? Yes, and dogs as well.

Mittelbergheim, Andlau. All around the ultimate peace, haze, labour; at Andlau there's a small weekly market. A stone fountain, the likes of which I've never seen before, is my resting place. The winegrowers subsidize everything here and are the backbone of these villages. In the church in Andlau, the priest is singing mass, a children's choir clustered around him, with otherwise just a few old women in attendance at the service. On a frieze outside, the most grotesque Romanesque sculptures. Holiday homes at the edge of town, all closed for the winter and shuttered up. But breaking into them would be easy nevertheless. A row of fish ponds there is exhausted, used up, overgrown with grass and brush. It runs along a brook.

A perfect morning; in perfect harmony with myself I'm walking briskly uphill. The potent thoughts of ski jumping make me feel light, like floating on air. Everywhere honey, beehives, and securely locked holiday homes throughout the valley. I chose the most beautiful one and contemplated breaking in then and there to stay the entire day, but it was too nice walking, so I walked. For once I didn't notice that I was walking, all the way up to the mountaintop forest I was absorbed in deep thought. Perfect clarity and freshness in the air, up further there's some snow. The tangerines make me completely euphoric.

Crossroads. Badly marked from here on. Naked woodland swathes with blue smoke all about from the woodcutters' campfires. As fresh as before and, like this morning, dew on the grass. Practically no cars up to now, and just half of the houses are inhabited. A jet-black wolfhound glared after me with his yellow eyes unflinchingly. When a rustling came from some flying leaves behind me I knew it was the dog, even though it was chained. All day long the most perfect solitude. A clear wind makes the trees up there rustle, the gaze travels very far. This is a season which has nothing to do with this world any more. Big flying reptiles soundlessly leave their vapour trails behind above me, heading directly west flying via Paris as my thoughts fly with them. So many dogs, from the car one doesn't notice them that much, the smell of the fires, too, the Sighing Trees. A shaved tree trunk is sweating water, again my shadow cowers far in front of me. Bruno flees, at night he breaks into an abandoned ski-lift station, it must be in November. He pulls the main lever for the cable car. All night long the ski-lift runs nonsensically, and the entire stretch is illuminated. In the morning the police seize Bruno. This is how the story must end.

Higher and higher, I've almost reached the snowline that begins at about 2600 feet, and then, further up, the border of the clouds. Foggy wetness begins, it grows dusky and the path ends. I inquire at a farmhouse, the farmer says yes, I have to go up through the snow and a beech forest, then I'd certainly come upon the road Le Champ du Feu. The snow is half melted, hardly any footprints, at last they stop altogether. The forest is foggy-wet, I know it will be unpleasant beyond that height. The farm was called Kaelberhuette, it's deathly still in the cloudy mist. It is impossible to know where you are here, only your direction is known. Although I've apparently attained the summit, when I don't reach the road, it strikes me as odd, and I stop in the dense woodland that finally consists of fir trees; thick fog has settled all around me. I try to work out where I made a mistake. There's no other solution possible than heading further west. As I pocket the map, it occurs to me that there's garbage strewn about in the woods, an empty can of motor oil and other things that could only have been thrown from cars. Then it turns out that the road lies only thirty metres away from me, but I can only see as far as twenty metres, and clearly, just a few steps. Following the road northward in the thickest fog I hit a strange circular outpost, with an observation tower in the centre resembling a lighthouse. Stormy winds, intense wet fog. I take out my storm cap and talk out loud, since all of this is barely believable after such a morning. Now and then I can see three white lines on the road in front of me, never any further, sometimes just the closest one. The big decision: follow the road north or south? It later turns out that both ways would have been correct, because I had been walking westward between the two little roads. One leads over Bellefosse to Fouday, the other down to Belmont. Steep slopes and slashing wind, empty ski-lifts. I can hardly see my hand before my face; this is no proverb, I can scarcely see it. Hath this brood of adders venom? Aye, and if downtrodden so have I. I yearned to kindle a fire; I would love nothing more than to see it already ablaze. T'would fill mine heart with dread lest thou break salt unto me. Meanwhile it's got stormy, the tattered fog even thicker, chasing across my path. Three people are sitting in a glassy tourist cafe between clouds and clouds, protected by glass from all sides. Since I don't see any waiters, it crosses my mind that corpses have been sitting there for weeks, statuesque. All this time the cafe has been unattended, for sure. Just how long have they been sitting here, petrified like this? Belmont, a Nothing of a province. The road was thirty-five hundred feet high, leading down now snake-like by a brook. Lumberjacks again, smoking fires again, then at 2300 feet the clouds suddenly blow away, yet below them a cheerless drizzle starts to fall. All is grey, devoid of people; downhill beside a damp forest. At Waldersbach no chance of breaking into anything, so I accelerate, to find somewhere in Fouday before nightfall. As there are hardly any possibilities even there, I decided to force open a tavern, that's locked on all sides, a big one in the centre of town between two inhabited houses. Then a woman came, didn't say a word and stared at me, so I didn't do it.

Outside town I go to eat at a transport café, and a young couple with something strange and oppressive lurking about them, as in a Western, enter the restaurant. At the next table a man has fallen asleep over his red wine, or is he faking sleep and lurking as well? The little duffel bag I've carried most of the time over my left shoulder, and which rests on my hip, has worn a fist-sized hole into the sweater

under my jacket. I've barely eaten anything all day, just tangerines, some chocolate, water from streams drunk in animal posture. The meal must be ready by
now; there will be rabbit and soup. At an airport, a mayor has been beheaded by a
helicopter as he was stepping off. A truck driver with lurking eyes, wearing worn-
out slippers, pulls out an extremely misshapen Gauloise and smokes it now without straightening it out. Because I'm so lonely, the stout waitress lends me an
inquisitive word over the lurking silence of the men. The exposed root of the
philodendron in the corner has sought tentative support in the radio loudspeaker
there. A small porcelain Indian figurine is also standing there, his right hand lifted
towards the sun, his left hand supporting the arm that's pointing up; it's a stately
little statue. In Strasbourg, films by Helvio Soto and Sanjines are showing two or
three years late, but showing nevertheless. Someone at a table near the counter is
called Kaspar. A word at last, a name!

Searching below Fouday for a place to spend the night, it was already densely
dark and damp and cold. My feet aren't working any more, either. I break into an
empty house, more by force than cunning, although another house that's inhabited is right near by. In this one workers seem to be repairing something. Outside
a storm is raging as I sit in the kitchen like an outlaw, burnt out, tired and drained
of all sense, because only here is there a wooden shutter that allows me to switch
on a little light without the glow escaping outside. I'll sleep in the nursery since it's
the best place to flee from in case somebody living here does in fact come home.
Most surely there will be workmen coming early in the morning, with the floors
and walls in some rooms being redone, and the workmen having left behind their
tools, shoes and jackets overnight. I get drunk on some wine that I bought at the
transport cafe. Out of sheer loneliness my voice wouldn't work so I merely
squeaked; I couldn't find the correct pitch for speaking and felt embarrassed. I
quickly split. Oh, what howling and whistling around the house, how the trees are
jeering. Tomorrow I have to get up very early, before the men arrive. In order to
wake up by the morning light, I'll have to leave the outside shutter open, which is
risky because the broken window will be visible. I've shaken the glass splinters
from the blanket; adjacent is a crib, plus toys and a chamber pot. All of this is
senseless beyond description. Let them find me here, sleeping in this bed, those
feeble-minded masons. How the wind outside is worrying the forest.

Around three o'clock I got up in the night and went out to the little porch. Outside there was a storm and heavy clouds, a mysterious and artificial sort of scen-
ery. Behind a stretch of countryside, the faint glow of Fouday was glimmering
strangely. A sense of utter absurdity. Is our Eisner still alive?

Wednesday 11 December

All l see in front of me is route. Suddenly, near the crest of a hill, I thought, there is a horseman, but when I moved in closer it was a tree; then I saw a sheep, and was uncertain as to whether or not it would turn out to be a bush, but it was a sheep, on the verge of dying. It died still and pathetically; I've never seen a sheep die before. I marched very swiftly on.

In Troyes there had been myriad clouds chasing through the morning dimness, as it started to rain. In the obscurity I went to the cathedral, and when I, the Gloomy One, had crept around it, I gave myself a push and went inside. Inside, it was still very dim; I stood silently in a forest of giants that had once gloomed ages ago. Outside, it stormed so fiercely that my poncho tore; I swung forward from bus stop to bus stop, seeking refuge in the covered shelters. Then I left the intolerable main route and walked parallel to it alongside the Seine. The region was very disconsolate, like outskirts that refuse to stop, interspersed with a few farmhouses. The electrical cables howled and swayed in the storm; I walked bent forward a bit to avoid being blown off my feet. The clouds were no higher than three hundred feet at most, just one big chase. Near a factory a guard screamed at me from behind, thinking I intended to enter the premises, but I was merely keeping away from the trucks carrying huge fountains on them. It's impossible to walk across the fields, everything's flooded and swampy. Yonder where the land is ploughed the soil is too ponderous. Fortified by the weather, it was easier to confront faces today. My fingers are so frozen that I can write only with a great deal of effort.

All at once driving snow, lightning, thunder and storm, everything at once, directly overhead, so suddenly that I was unable to find refuge again and tried instead to let the mess pass over me, leaning against the wall of a house, half-way protected from the wind. Immediately to my right at the corner of the house, a fanatical wolfhound stuck his head through the garden fence, baring his teeth at me. Within minutes a layer of water and snow was lying hand-deep on the street, and a truck splashed me with everything that was lying there. Shortly afterwards, the sun came out for a few seconds, then a torrential rainfall. I grappled forward from cover to cover. At the village school in Savieres, I debated whether I should drive to Paris, seeing some sense in that. But getting so far on foot and then driving? Better to live out this senselessness, if that's what this is, to the very end. St. Mesmin, les Gres. I didn't quite make it to les Gres, due to my escapes from the enormous black wall racing towards me. I broke into the laundry room of an inhabited house, noticed by no one. For five minutes the Infernal reigned outside. Amid thick lumps of hail blowing horizontally outside, birds were fighting. In just a few minutes the whole

Thing swept over me, leaving everything white, an unsteady sun twitching in pain afterwards and, beyond it, deep and black and threatening, the next wall approached. In les Gres, shaken to the soles, I ordered a café au lait. Two policemen on motorcycles, wearing rubber uniforms that made them look like deep sea divers, likewise sought shelter. Walking doesn't work any more. The blizzard made me laugh so hard that my face was all contorted when I entered the cafe. I was fearful the police would seize me, and so in front of a bathroom mirror once more I quickly made sure that I still looked fairly human. My hands are slowly getting warm again.

Walked a long way, a long way. Far off in the open fields, when another one of those stormy fits came and there was no shelter near or far, a car stopped and took me a little way to Romilly. Then onward. I stood leaning against a house, directly under a window, during a hailstorm, as again there was nothing better to be found in such haste, and there was an old man inside—so close I could have grabbed him with my arm—who was reading a book by the light of a table lamp. He didn't realize anything was pouring down outside, nor did he see me standing near by, breathing on his window pane. My face, assessing it in a mirror again since an idea was stirring, wasn't altogether known to me any more. I could swim the rest of the way. Why not swim along the Seine? I swam with a group of people who fled from New Zealand to Australia, in fact I swam in front, being the only one who knew the route already. The only chance the refugees had of escape was to swim; the distance, however, was fifty miles. I advised people to take plastic footballs with them as additional swimming aids. For those who drowned, the undertaking became legendary before it even began. After several days we reached a town in Australia; I was the first one to come ashore, and those who followed were preceded by their wristwatches, which drifted in half underwater. I grabbed the watches and pulled the swimmers ashore. Great, pathetic scenes of brotherliness ensued on shore. Sylvie le Clezio was the only one amongst them whom I knew. When it started to rain very hard again, I wanted to seek shelter in a roofed bus stop, but there were already several people there. I hesitated before finally creeping over to a school for cover. The gate which served as an entrance for cars closed shut, making some noise, and the teacher eyed me from the classroom. At last he came outside in sandals and blue overalls and invited me into the classroom, but the worst was over by then, and I was too much into the rhythm of walking to be able to rest very long. The distances I cover now are quite large. When I left I replaced the iron gate in its lock very gently, so that I left without further ado. Walking endlessly up to Provins, I decided to eat prodigiously, but a salad is all I can get down. When I have to get up now, a mammoth will arise.

Saturday 14 December

As afterthought just this: I went to Madame Eisner, she was still tired and marked by her illness. Someone must have told her on the phone that I had come on foot, I didn't want to mention it. I was embarrassed and placed my smarting legs up on a second armchair, which she pushed over to me. In the embarrassment a thought passed through my head, and, since the situation was strange anyway, I told it to her. Together, I said, we shall boil fire and stop fish. Then she looked at me and smiled very delicately, and since she knew that I was someone on foot and therefore unprotected, she understood me. For one splendid, fleeting moment something mellow flowed through my deadly tired body. I said to her, open the window, from these last days onward I can fly.

Thin Air on Mt. Audubon

Andrew Schelling

Don't slip on the spruce brachiopod wracks or hidden ice breaks. At this altitude
a mica of volume cuts the Russian tundra into consonants. Filled with dwarf
trees, never too late, the year's squall exposed on a barometric edge. At final
ridge lift, some went melodic, some danced, some went into sculpture, most
drew close with a shrinkage of hide. When the glacier returns the rock scrapes
copper.

> Food chain stuff
> food utility sacks—
> sun going hard west.

Today the juniper glitter tomorrow the scree. A bare push south would drop you
through that precipitous col. Get back or get bleached, like the photo of Mallory
on Chumolungma, fallen and frosting the yellowish lichen patch with skin gone
porcelain under the wind. Was that when your thumbs went funny? Metal trouser
buttons more icy than a snow goddess glance, a frosted nipple, a shock of water.
The wind strings you like jerky. High on Audubon, low on oxygen, there
among singing weathered peaks, the primal taste came into your mouth as a few
times on mushrooms. First that inward darkness. Bitter, salt, meat, hormonal.
Distinction flavored with rock. What rock? With granite sediment. With
chimney. What chimney?

> Born with a shock,
> one blink and it's gone.
> Hand me that boulder.

To study the wind unobserved, to write snow banners, summits of where. Fox color hogback's a tectonic ripple below. Pine's absent, fir a kerosene tint. And green? Nowhere green. Taste was always more basic than touch, you discovered the flavor in marsupial time. Sentient, feeding on, loving nerved flesh. Mostly love. The first taste of a woman was shocking—that recognition—ingestion—left you cuntstruck? Back so far you were bent on the northward slope, splitting the outer husk. Thin air is where you learnt the uncertain answer. The riddle is what did you cast at birth. Taste of her inward self? Vouchsafed? Salt meat taste of your own. You never got over it.

> Dear dark mammal,
> clutching my life ever since—
> who amid ice, stone?

Ellen Scott, *Self-Portrait in Air*

From *The Substance of Forgetting*

Kristjana Gunnars

Perhaps we already recognized that a million significant perceptions had passed in the air. That we define ourselves in terms of each other. That we cannot exist in a monologue. Yet our dialogue is without words. There is no official agreement, no absolute truth. It is only an outdated morality that says I cannot desire whomever I want. The morality of a world I never saw and never lived in. A remnant of a past my forefathers have told me about but of no concern to me. In this world there is no morality. No truth about human actions. The only truth I find is what is not expected. Not accepted. Not officially correct. My mistakes are the most significant thing about me. My mistakes define me. What I cannot say defines me. What I cannot say to him because he cannot hear defines me. I have known him for a long time and this is our first meeting. What was it that passed in the air before? For this reason perhaps we lagged behind. We were going from one place to another. It is immaterial where we were going except we had to go. It is immaterial who was with us. People. Several people walking ahead of us, going to an elevator to be lifted up. We must have been in the bowels of an institution, deep below ground in the caverns of a building. Jules made us lag behind until all the people had rounded the corners ahead of us and we were alone. He was laughing and I was laughing and I no longer remember why. There is a point at which the inter-stated dialogue becomes an official story. A point at which others can recognize that something of significance has happened. That is where Jules becomes another. He becomes a man in a green coat whose arm was around me. The man was laughing at something. He was pulling me back so the others would disappear.

His arm was around me and he stopped to pull me toward him and he kissed me. He had never kissed me before. There was no reason for it. Then he looked at me as if he had surprised himself. Just as though a blue haze had materialized from the floor and the genie had swung out of Aladdin's lamp. Something extraordinary. Something said without words. A kiss. But a kiss is significant. It tells an official story. It is something identifiable to others. It is where we agreed. It is so much easier now to talk about desire. Perhaps it is just his presence, one spirit desiring another, one language desiring the presence of another.

The dance of approach and rejection is stilled in the Okanagan Valley. In the early morning only a mild haze penetrates the air. The water is almost blank. Sometimes the sun that just creeps over the hill shines a mellow band onto the incandescent water. Then ripples can be seen. A tiny breaking of the surface, a ruffling of the tranquility of dawn.

It is a landscape that requires no stance, no answers. No position, no preparations. A landscape that couches those of us who live here in mellow arms and asks nothing of us. A gentle mother who allows us to rest awhile. We can rest just by looking at the hazy green morning of early spring. When all is awakening.

What is a sentence that is at rest? Could I write a sentence that has no tension in it? No elliptical curve from desire to return? A sentence that acknowledges it is tired and wants to rest. The sentence knows what it wants. To rest awhile. A sentence without an Other. Without a lover. Without a desired object. Perhaps even a sentence without a subject. No ego. No narcissistic ego settling its image over the world.

I came to this valley because I wanted all that tension to seep out of the phrases all around me. To uncharge the battery of my language. I was tired. The words were crammed too full. They could not hold the wealth of information and counterinformation I had put there. They were so full that it was impossible to recognize what was in them. I wanted to see the disappointments that had accrued fall like fluff from the branches. To see the naked branches.

As the morning lingers the haze melts away. I see the water more clearly. It is blank. There is no wind. The pastures and orchards are reflected in the lake. The sun is behind a thin cloud and the air is soft. So soft. It has the sheen of silk and the texture of cotton.

It occurs to me I have been dreaming. That Jules' windblown face is a dream. His dark hair ruffled by the chilly winter wind. His collar turned up against the cold.

Press clippings flying all over the railroad tracks. Cookies falling out of boxcars and crumbling in the last bit of snow on the ground. A dream we both awoke from. It was a new day. The palm of his hand lay across his forehead. He awoke in a large bed with pillows and sheets. There was an opening between the curtains where the bright light blasted in. He was thinking, his elbow in the air and his hand on his forehead.

"Where are you?" I asked and could barely hear myself asking.

"Have you heard of cloud nine?"

"Yes."

"I am on cloud twelve. Cloud fourteen."

All the clouds were rising. It was easy to see the clouds lift lightly into the air. One by one they rose from the sleeping valley. From between the hills where their own weight had pulled them down. The sun was beaming on them and the water that was in them was dissipating. Their sorrow was vanishing. As they grew lighter they floated up. Behind them was blue sky, so azure it was like a gem. The clouds floated into the sky, becoming whiter and lighter as they did so. They were flying home. I could no longer count the clouds, there were so many. But when I looked out the window I saw the angry snow and I knew we were not dreaming. That rising from this bed would be the beginning of a distance. An immense distance we would be unable to close again. Wildly different lives waited to greet us at the ends of that opening. A difference of French and English. Of city and country. A distance of mountains, lakes, rivers, fields, pastures, suburbs. Everything that lay across the country would lie between us. Time itself would lie between us. We would forget each other. Time would place our images in a dream. It is a dream you did not know you had. I was thinking that even though his arms are around me and his body is around me I can feel the unbridgeable gap. The separation is in the moment even while the moment is expanding to contain all time in it. The waters are rushing to the edges of the lake. The lake is rising and overflowing its boundaries. The floating bridge is under water. It is impossible to travel across.

How quickly the clouds dissipate. When I arise in the early morning we are enveloped in thick cloud and fog. It is as though we have been moving during our sleep from our comfortable beds with many duvets and cushions billowing about us into an ethereal sphere high above the earth. A place of nothing but cloud. When I look out the window I can see nothing at all. There is only a thick gray substance, as if the whole cottage had been placed inside the center of the fog.

Then the day advances and the fog begins to separate. The thick haze over the lake remains, cradling the water. Above, the mist releases itself and turns into billowy white clouds. The fluffy bundles start floating up and between them and the lake the image of mountainsides appears. Soon I can make out the slanting hillsides, the open pastures, the forest. There is still a bit of snow on top of the mountains. Between the trees I can see layers of white dust.

The mist over the lane itself begins to rise. Underneath I see how the water is perfectly blank. In its face I see reflected the rising clouds. The water appears like a giant layer of marble. I know if I went down there and looked in the blank surface of the lane I would see my face clearly. I do not know if it would be the face of a stranger.

It has been raining all night. I could hear the heavy drops falling in the eavestroughs and on the roof. The rain fell straight down because there was no wind. There was just the sheer force of gravity bringing down the water in the clouds. In the morning puddles lie still on the porch. The front steps are wet. Huge drops still hang from the roof, uncertain about letting go. The grass holds up beads of rainwater for the sun to see.

I put on my coat and boots and walk up the wet gravel of the road. The path from my cottage goes through the forest. The forest on such mornings is somber. The overcast sky keeps the pine trees dark and moody. In the hills above, the trees appear to be waiting for the sun. Like all the world they wait for the warmth of sunshine. I like to walk among the ponderosa pines. Everywhere the gentle ponderosas stand with tufts of needles on their branches.

There was a sudden blast of wind in the night. It came and went so quickly that it was hardly detectable. But it was so strong it tore objects from their places. When I come up on the road I see that a giant jack pine has been ripped out of the earth by the wind. The trunk lies across the road in complete defeat. The branches have splintered off in the violence of the impact. Bits of branches lie everywhere. When I look at the earth where the tree was uprooted I see the roots spreading dizzyingly into the air. I can tell how shallow the roots were. I see it is no wonder the tree fell down. Perhaps this sandy soil is hostile to such trees. Perhaps the whole forest is tottering on the brink of collapse, the tiny root systems just barely holding the heavy trunks steady.

I cannot help wondering if it is so hard to find a foothold in the West. Not even the native trees can do it. A little blast of wind is all it takes. Perhaps this is not true. Perhaps this is not true at all.

I am the place from which the voice is heard, Jacques Lacan said. *This place is called Jouissance.* There is nothing here to be seen. It is the empty space where the breath breaks through. It is a wind in a tunnel, strong enough to fell whole trees with shallow root systems. There is no Other for this *jouissance* unless it be through desire. I cannot prove that the Other exists unless it be by loving him.

We are trapped in images not of our choosing. We have come a long way only to find that we went nowhere. We came all this way from the apple tree only to find another apple tree. The apples here are falling before they can be picked. It is because the tree has been neglected and the worms have feasted on the fruit. The apples lying on the ground are riddled with holes. They have been rotting in the grass and the bees have found them.

In the valley where I live the fruit farmers complain about their apple orchards. They say apple trees are the worst. Apple trees attract infestations. They have to be sprayed every season. In the early summer you can see sprayers in the orchards. They are people who appear to have landed in a spaceship. They are outfitted in protective clothing. Their faces covered in masks. They make you think of the war in the Persian Gulf when everyone put gas masks on. In their hands they hold devices from which a hard poisonous spray covers the fruit.

The world has a device it can use to hold me hostage. It is a device of organic symbols where they say words are coming from. In my joy I am illicit. My joy itself is forbidden because no document has sanctioned it. The universe is a defect. I myself am a defect. The cosmos would be purer without me. And that is what I am trying to say. I am trying to say it is the defect that matters. You know you exist because there is a break, a problem. Because something has been forbidden and it is so hard to believe.

Laughter itself is the breath of the mistake. A defection from the rule of tyrants. Because there were iron curtains everywhere and you saw a break in them where the daylight filtered in. In the blue light of dawn you saw clearly the man lying beside you. He was sleeping and you saw whom you had been with. Suddenly there were no tyrants. There were no dictators and the curtain had broken. The veil had been torn and in the tearing of the veil you saw his face imprinted. The face was clearly there. A shadow cast into your world and you were laughing.

Because when you arise in the early dawn you recognize that the unspeakable has laid hands on you. In the unspeakable there has been a blessing. A gift in the palm of his hand. He transferred the gift to you when he placed the palm of his hand on your body. A gift of language. Of accidental joy.

You want to say you are not trapped in these images of valleys and mountains, of canyons and peaks. Of streams and bridges, of towers and ruptures and spiral staircases. That like a circling hawk you can fly beyond them all. They cannot hold you.

It will be much easier now to see the airplane land in the valley. To watch my bags coming out a trapdoor in the wall, floating forward on a conveyor belt. To drive to my cottage and see that the ice on the lake has melted. The ice fishermen are gone. The ducks are swimming freely. A boat is on the water. The iron hold of winter has loosened. The bars are falling off. Their joints are weakened and they collapse at a touch.

I would say I am tired. Winter has been long and spring has been slow. I no longer know what I have to do. Work that is made for me piles up. I have confused it with work I want to do.

I have begun to stare at the lights on the mountain at night. On the other side of the lake lights are scattered where people live. The lights are reflected in the lake. All the ice has melted and strings of light can be seen on the water. When the moon is full there is a wide band of moonlight on the water as well. In the country the air is clear and all the stars are visible. The lights of the lake and the lights of the mountain are met by the lights in the sky. Everywhere the world sparkles at night, like the sun sparkling itself in the water during the day.

From my cottage I see shooting stars. They pass overhead at great speed, going from south to north. They are moments flashing across the night sky. Then they are gone.

In the spring, before the leaves are on the trees, the early mornings are noisy. Hundreds of different birds are calling out the daylight hours. I do not know what half of them are. There are more birds than I imagined. They are delighted at the prospect of blossoming orchards and clear blue lake water and tiny red leaves on little plum trees. The grass has become bright green. The blue haze has settled over everything.

I am thinking home is where you choose to forget and choose to remember at the same time. Nothing hinders your choices. Nothing forces you to remember and nothing forces you to forget. There is no reason to repress any memory. There is no reason to hold it up against the daylight either.

III

Taking It In, Letting It Out

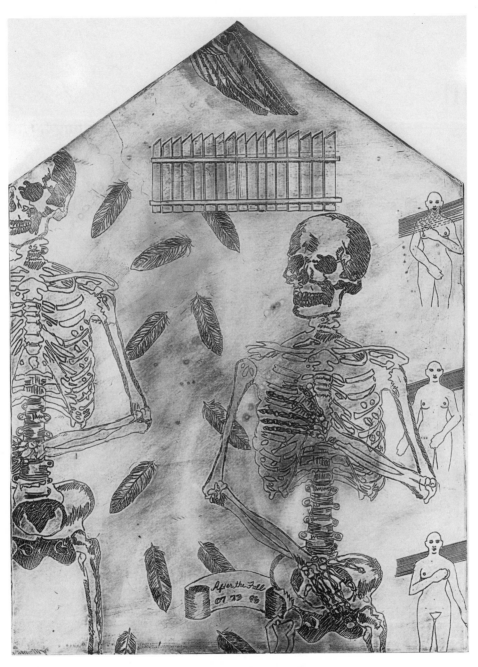

Louise Weinberg, *After the Fall*, 1996. Ink and oil on copper and wood.

Notes on Emphysema

Hayden Carruth

1. Smudgie, my beautiful white cat, lies curled on the bed beside me. She doesn't know she's breathing.

☒

2. At the beginning of the instant of oxygen deprivation, as when one exhales as far as one strenuously can and the lungs ache, a diminutive twitch and tingle are felt in the little finger of each hand, as if it were a mild electric shock. The brain sends out its small distress signal. You had better pay attention.

☒

3. Breathing is called an autonomic activity, and of course it is. Like the cat, a child doesn't know it's breathing. Yet sometimes in the duress you come to know too well you must concentrate every consciousness you have on one more breath.

☒

4. Climb the stairs. Lie down. Stare at the water stain on the ceiling, at utter familiarity. Slow your heart and still the gasping. Gradually, gradually. Notice the rise and fall of your diaphragm, the diminishing pressure against your stomach. After five minutes take one deep breath. A great exhaustive sigh. Now fall asleep.

☒

5. The messy condensation inside the oxygen mask makes you think of a suckling infant. All that moistness on the brink of suffocation.

☒

6. After walking up to the brush pile with a handful of winter twigs and small branches, do not catch your breath standing up. Fall to the ground. The grass. Turn on your back and cover your eyes with your arm. Open your mouth wide. Try not to choke.

☒

7. When you increase your use of the medicated inhalers to three times a day, four times a day, commit yourself to the pain of knowing that this will not help you.

☒

8. Commit yourself to pain.

☒

9. Remember frequently what you have read, heard, and seen on film. The end will be a plasticine tent in a white room. Vague faces, unintelligible voices. The thump and rasp of the breathing machine, like a pile driver beside your head. A little fire of burning tinder in the bottom of your throat.

☒

10. How alone can you get?

☒

11. "Now I am dying. Now. I am afraid. Please give me a cigarette."

☒

12. The x-ray shows nothing. A diseased alveolus looks the same as a healthy one. But the doctor can tell anyway, because the lungs grow grotesquely upward, squeezing into the collarbones, in order to get more air. The whole lungs, top to bottom, will no longer fit on an x-ray plate.

☒

13. See the basketball player. Sprewell of the Knicks, for instance. Whirling like a tempest from one end of the court to the other and back again, never ceasing, and then the shots, the lay-ups, baseline jumpers, perimeter bombs, and the steals, the breakaways, dazzling, all magical and dazzling. But the really remarkable thing is how that man can breathe.

☒

14. Learning to cough. Not the shallow, sharp spasm of earlier years. It must come from a deep place, as deep as you can get. Let it be slow, somewhat prolonged, yet very vigorous, a great huff, like blowing out many candles. Mouth wide open, trying not to wheeze. Bring up that awful, heavy clot of phlegm. Put it in a tissue. Examine it carefully for signs of blood.

☒

15. The endless questions. Why do you smoke? Is it addiction to nicotine? Clearly not. What did nicotine ever do for you? Nothing, nothing at all. There is a difference, an important difference, between addiction and compulsion, no matter what the doctors and other innocents may think. Emphasize this. But where does the compulsion come from? From far in the back of the cave where the firelight doesn't reach and you are a boy alone. And why is it so strong? That's obvious. Because it saved your life.

☒

16. Often of a summer evening one feels a chill, a shiver. Is it the dreaded virus? The virus which is nemesis for emphysema, especially when the emphysema is combined with chronic bronchitis, as it usually is. A touch of pneumonia can bring on the ultimate misery in no time. Ultimate, meaning what it means. Take

aspirin, take whiskey. Go to bed, snuggle down, try to sleep in spite of a raging brain. Hope the chill is only an old man's fragility. Hope for the best.

⊠

17. If you look in a mirror and see the lineaments of hope, smash the mirror. That's what Paul said, long ago.

⊠

18. Is the emphysema more deranged than the other demons who have dwelt in one's being from the start? Probably not. But it does have a material presence that the world can see. It has more power.

⊠

19. Good morning, emphysema. Good afternoon. Good night. Did you have a good day?

⊠

20. Remember what it meant to "take a walk"? Along the brook, through the hemlocks and spruces, around the bog, across the goldenrod and hardhack to the bank of fireweed, under the knoll where the purple-fringed orchis grew, wandering the hardwoods with the squirrels and ovenbirds scolding, up to the top of Marshall's big rock? Breathing the clean air?

⊠

21. You can't budge? Unh. Is it the emphysema? You bet. What does it do? Gives me the granddaddy of all depressions. What can I do? Give me a cigarette and go away.

⊠

22. Would you like something to drink? Thanks. I'll have a little Kentucky cyanide.

⊠

23. First thing in the morning you can breathe—a little—and you can move—a little. It hurts. Five-thirty A.M. with gray light at the window, probably another day is beginning. Yeah. But man, like who gives a rat's ass?

⊠

24. He would like to see his son, say a few words, hear his voice, touch him on the shoulder, he would like this more than any words could say in any language. But the boy lives five hundred yards up the hill.

⊠

25. Choking and gasping in the night. The woman comes, the belovëd. She who is so constant, beautiful, and smart. But no one is smart enough to know what to do about this. Moments of overwhelming mutual disappointment.

⊠

26. When friends come you sit immobile, smiling, breathing quietly, the way the Buddha sat. But everyone is aware of the difference.

⊠

27. The Angelus. One doesn't know what it means, being resolute in unchurchliness. But the word has such a pleasant sound. One would like to go to sleep by—at, on, with—the angelus. And perhaps never wake up.

⊠

28. After all, which means after a very great deal, perhaps a Catholic hospital would be best. He has never known a woman he could call sister.

⊠

29. When he is gasping and coughing, when arthritic pain rages in his back, shoulders, and hips, the dog does not understand. She will bring him her ball, the

blue ball with a hole through it, and demand that they go out to play. Then she becomes visibly reproachful when he is unable to respond. He never knows whether to be sorry or pissed, and in fact he is both—at the same time uncomfortably pissed and sorry—for the dumb beast's sake, for the man's sake, and for the world's.

☒

30. What day is it? Monday, apparently. What date? The first of May. But it could be next Christmas, or possibly even the Christmas after that. The change of seasons is no more than a change of wallpaper now.

☒

31. You reflect that emphysema affords one benefit after all. Your whole life you've suffered from acute acrophobia, sometimes dangerously. Jump, jump, the demon shouted, when you stood at the top of the precipice on Capri. But now you can't climb high enough for the vertigo to set in. No doubt there are others.

☒

32. When did one draw one's last normal breath? Probably in sleep. When it did no good. Otherwise every breath, even in repose, is taken too quickly and expelled too forcibly, often through half pursed lips. As if you were trying to whistle long after you'd forgotten how.

☒

33. What is a lung? Well, now that you ask, a lung is a sponge. And you know what happens to your kitchen sponge after you've used it a couple of years.

☒

34. The computer doesn't breathe, but sometimes it sighs. Especially when you turn it off.

☒

35. "Get rid of the gasper. Why don't you smoke one of those fine cigars from Philadelphia?" "Why don't you mind your own fucking business?"

☒

36. The beginning. The ending. Inexorable. You used to think disacquiescence was a virtue that somehow lent dignity and integrity to human consciousness caught in the bind of essential injustice. Why not now? Well, that's the kind of an idea one had during the middle.

☒

37. You have two inhalers, called "puffers" in the trade, which for two years you've been using as prescribed, two inhalations of each twice a day. You use an air chamber so that the maximum amount of medicine will reach your lungs. After each puff you hold the medicated mist in your lungs for thirty seconds if you can, though you probably can't: in mid-case emphysema twenty seconds would be more realistic. Between puffs you exhale until you're on the brink of fainting, in order to clear out as much of the smoke and stale air as possible. The whole process takes twelve minutes, morning and night, and produces no noticeable effect. On the other hand, if you neglect the puffers for a day you become aware of increased pain and greater shortness of breath. The strange thing is that even though this exercise is directly related to your hope of staying alive, it becomes a colossal bore. You have to flog yourself to it every morning and night. Is it that staying alive is something of a bore too? In addition you have a third puffer for use in emergencies, more powerful and with more evident side effects. You use it just before you dial "911."

☒

38. Ten times a day: "Why don't you quit smoking?" One has no hope of penetrating the so-called minds of these people. Talk about exclusion! Talk about discrimination!

☒

39. He is a professional. He has read at least ten poems a day for sixty years—that's 220,000 poems and probably a lot more. Not one of them was about breathing.

☒

40. Do not hold your breath when doing something difficult, such as walking quietly in order not to disturb the sleeper or scraping the old inspection decal off your windshield. This seems to be a natural human practice, but it isn't necessary and it only makes matters worse.

☒

41. According to Victoria Maxx the Tletuecans had only one god, whose sole function was to take upon himself all the ills of the people. But it's hard to imagine a divine person with emphysema.

☒

42. Even so you relish what you still can do. You heave the bag of trash-paper—all those photocopied manuscripts, which weigh 100 pounds—into the back of your little station wagon, and then lean against the wall of the garage until you recover. Then you drive fifteen miles to the landfill, the recycling center, slowly through the beautiful tracts of woodland, accelerating to 75 through the flat cornfields, to feel again the thrill of speed. And you are immensely gratified by your own civic virtue. You drive with a calm heart and even breath.

☒

43. Oh, to have a house with an elevator, two bathrooms, and a mechanical bed!

☒

44. Your friend the psychiatrist told you that the reason you fall asleep so easily is because what you are doing is not interesting. Is it helpful to know that nothing is interesting? If only one could dream of interesting things! Instead, dream after dream, the same old inconsequential religious nonsense.

☒

45. Caught in the irreversible, you do things that you absolutely KNOW are futile. Yet you do them. Again and again. To what a nonsense of misery the human spirit can be reduced!

☒

46. *Breath* → *Braeth,* Old English for "odor," related to Greek *atmos,* "vapor." Not much help there. Latin *spiritus,* meaning both "breath" and "spirit," whence "respiration," "aspiration," "inspiration," and of course "expiration."

☒

47. What does a square with a cross inside it mean? An expansive inner tension within an inflexible confining limit. In other words, breath.

☒

48. In the computer it is called a wingding . . .

☐

Tuula Närhinen, *Anemographic*. Drawing made by wind and birch, part 1

The Laugh

David Appelbaum

The laugh rises to the top. It is a bubble driven heavenward by a physics of effervescence, not to be denied by all obstacles of gravity. Neither misfortune, sorrow, tears, nor pain slows it for long, for the laugh is light and in an upward direction seeks the being of light. It rises to the occasion, any occasion, even death.

From over a crest of the hill, from the glen, comes the sound of children at play. Shrieks and shouts, ballyhoos and singsongs, are ambiguous with respect to suffering or joy. They are loudest of sounds, but do not say whether concord or discord rules the hour. Only with closer listening do sounds—"hidden excitedly, containing laughter" (Eliot)—resolve the question. The laugh is conferred by harmony, and its spirit is of relatedness and ease. All sides are brought back to balance by laughter that rises like a carnival tent over the treetops.

The laugh suffers a great many impersonations that becloud a simplicity of spirit. Each appropriates the laugh and qualifies its unconditional freedom for a limited end. Cruel and taunting laughter, mocking and derisive laughter, forced and contrived laughter, bestial and tormented laughter, precious and lewd laughter—all strive to make use of choiceless ease. The laugh, when pure, completely pierces a crust of ego. Below the stiffness, what has no use determines the course of things and enables use of this or that. Like a subterranean fountain, the laugh that emerges in sunlight is subject to conditions of the depths. It can be controlled according to fashion but never without revealing its unfashionable origin. Its unlearned note, moreover, can be learned—that is, copied—to produce effects of fear, anxiety, doubt, greed, hesitancy, the whole scale of human emotional life.

To sing a song of the laugh is quite different from being sung by it. A child knows the difference.

Of the everyday spirits, the laugh is unique in being an *arche*-sound. Two other *arche*-sounds are the word one calls oneself by—*I*—and *Om,* "That syllable indeed is the supreme" (*Katha Upanishad* 2.16). The three together divide and exhaust a ground of being, in the way the good, the true, and the beautiful do. *Om* expresses goodness, *I,* truth, while we, hearing strains of laughter rippling an evening calm, know the laugh as beauty itself. Such sounds are uncreated. They issue forth from wholeness like sympathetic vibration does from a tuning fork. Each is a smaller whole resounding in like pitch with the greater whole. Each has a power of evoking the same sound in another. We are unsurprised at the contagiousness of the laugh. We are, however, astonished at how the laugh transmutes repulsive, ugly, loathsome, or ungainly looks to beauty. As a forbidding mask drops, the grace of unpretentious laughter opens the eyes to what waits excitedly hidden.

A bubble of joy, the laugh rises to the top "like the bloom of youth does on those in the flower of their age" (Aristotle, *Nicomachean Ethics* 1174b.33). It lives in its own time, and when it comes, it stays the moment a little while. In the fairy tale, all who tried were unable to make the dour princess laugh. Only when she by chance saw a fool who carried his donkey on his back did she burst (like a bubble) into laughter. Ice melted, and levity easily escaped into the atmosphere. Means exist, moreover, by which to prod, poke, tickle, and eventually prick the skin into a laugh. They also serve this dynamic spirit in the way in which a cultivated rose serves the same beauty as a wild one, but differently. Approaches to laughter are three: intoxicants, jokes, and comedy.

Wine returns one to unpretentiousness. The grape is of earth and like many earthly things has a tender pulp protected by a tough skin. When they are fermented together, one grape is no longer separate from another. Together, their substance is transformed. The action of wine on a psyche is analogous. Integument softens and grows permeable. As one's boundaries—limits of a pretender's domain—are no longer sharply defined, one blends more easily with an element that all have in common. Communication expands beyond a narrow, pent-up, mental channel to embrace organic feeling, including a feel for wholeness. In a Dionysiac or bacchanal, release is often explosive, possessing an effervescent force of the grape fermented in the bottle. Into the melee, the laugh escapes.

No longer pretending to be different, one joins together in laughter. Riotous, inarticulate, sexual, the laugh of revelry returns one to a natural past in which impulse and instinct command action. Such a wholehearted surrender to absurdity cleanses a mind and sensitizes it to forces beyond human dwelling. The risk is great. Vision can pierce the skin of things and swell to cosmic proportion. Ecstasy, however, may lead to a rapture that abjures the acid soil from which the grape grows. Then mind is possessed—or dispossessed of its ground. Excess and frenzy may fill a void left by temperance. An encounter with energies of the intermediate realm may prove fatal. One could die in a mad fit of laughter.

Still, wine need not abandon balance but might weigh things more urgently. Sobriety then yields to, as Alcibiades says, "philosophical frenzy" or "sacred rage" (*Symposium* 218b). Temperance ceases to hide truth behind measured phrases. In the openness, the laugh takes in one's own idiocy. A bubble of conceit pops. Bang! Or, in Alcibiades' image, a paper satyr (an ancient form of the piñata) bursts and spills its penny contents. In the wake rise real questions. What attitude to have toward the stupor encasing one's life? Whence the impulse to meet one's destiny? How to traverse the space between heaven and earth? As laughter precedes interrogation, it must follow it. When laughter dies, one takes oneself too seriously. An inner beauty—the source of the outer—is forsaken for a mind's vanity. Thus Alcibiades asks, "Are you asleep, Socrates?" "No, I'm not," comes the answer (*Symposium* 218c).

Which brings up the joke. Wine provokes the laugh by unstiffening the carriage of ourselves and revealing our graceless, pompous, hysterical manner. It is a lubricant of earth and reinjects an earthiness into our flights of fancy. Its effect is organic. The intoxicated laugh takes one to midregions between shores before succumbing to inebriation. A good joke, by contrast, works on a mind, since it conveys the fact of timeless intelligence. "Eternity is just time enough for a joke" (Hesse). Logic steers mutable mind. It directs thinking along time-worn channels, lest thought get other ideas and wander distracted into the unknown. It thereby safeguards soundness, sanity, and sobriety. The joke is a rebel against convention. It is daring, daunting, and dashing to cherished ideals. Its content is the same no matter what its disguise. It is a perception of contradiction. Logic asserts that something and its negation cannot both coexist. The joke says, they do, and by the way, they're in you. By our contradictions are we known. They expose us, lies exploded, fakery thrown to the wind, air let out of the balloon. Helium going skyward is none other than laughter, traversing the region toward the sun.

Put more geometrically, the joke is a recognition of incongruity. The right hand is in the wrong glove, or the shoe that is supposed to fit does not. The road curved left while the convoy went straight, straight into a quagmire. The disagreement is between mind and reality. The two cannot be made to correspond. The joke turns on our insistence that we are right, no matter what the facts are. In its clearer light, our very limitations, dissolved, reveal the way to greater freedom.

> Mulla Nasr Eddin's friend wanted to borrow his clothesline. "Sorry," said the Mulla. "I'm using it. To dry salt." "How's it possible to dry salt on a clothesline?" "It's easy when you don't want to lend it."
>
> The watchman caught Nasr Eddin prying up the window of his own house at midnight. "What are you doing, Mulla?" he asked. "Locked out?" "Shhh. People say I sleepwalk. I'm trying to surprise myself and find out."

The joke brings the laugh by revealing our laughable attempts at reason. That is the joke's mission, and it is taken most seriously. Surprisingly, by voice and posture, the joke is akin to a beggar. The original Sanskrit word for *joke* means "he implores." The joke begs our attention for a moment, this moment, to remind us of a gap in our logic. Calculation again has fallen shy of the mark, deduction is flawed, and (by the way) the choice of means was horrendous. The joke beseeches that we stop. The situation needs another look. Our eyes are trained too narrowly. The joke prays we forget our self-importance. We have taken hold of an idea of who we are, from which all reasons follow. To release our hold, it offers us release to the laugh.

The joke begs our alms whose coin is laughter—silvery, throaty, or visceral. With comedy, third provocateur of the laugh, we pay in a different way. Witness to a theatrical drama, we are shown a humorous edge to things. The edge cuts to the quick with trenchant satire, sarcasm, and irony. Human folly thus dissected by ridicule or scorn is seen to be what it is—a weakness that can be made into strength. We are invited to laugh because we are transparently not what we seem to be. Laughter, in rising, separates us from ourselves. A joy of nonattachment restores resolve. In the laugh is consciousness of humanity and of that from which humanity springs. Great is comedy's means of evoking the *arche*-sound, and great is the laugh's power to lead us across the intermediate realm.

A laugh begins in silence before taking off as an audible vibration. But what of the laugh that never grows audible? The one that stays so close to the ground that it blends perfectly with other sounds of earth? Then we have the smile, a soundless laugh. The smile is itself a miracle. Both words stem from a common root, the Latin *mirari* (to wonder at). Filled with wonder, one is emptied of preoccupation. Struck by astonishment, one stops perceiving through expectations and confronts what is. Marvel and surprise mix, dissolving preconceptions that clog the senses. The moment is an awakening of intelligence. Its natural expression is the smile.

According to legendary accounts, by the miraculous spirit of the laugh the Buddha conveyed his understanding. Sakyamuni was discussing his ideas when he arrived at a critical point. Finding no words, he lifted a single flower. All around him were blank looks except that of an old man, Mahakasyapa. The old man smiled, no more. His smile signified transmission of wonder. The Buddha perceived his awakening and pronounced Mahakasyapa his true successor.

The smile—is it not a return to an unlearned wonderment? An infant's smile spreads across her face with the bliss of existence. "The smile that flickers on baby's lips when he sleeps—does anybody know where it was born?" (Tagore). In the infancy of consciousness, in deep, dreamless sleep, the eternal mother reigns. The immutable realm is hers. Clothes and all ephemeral things with which the baby is dressed belong to the father. Sleeping, an infant becomes naked again. It rests beside the mother's breast, where its repose lacks no thing. All is sufficient, all is plenty. It is nourished simply by being where it has come to, unsapped by dream, perception, or attainment. There is wonder at a perfection of support, nothing more. From awareness, like a pocket of oxygen trying to escape the depths, an infant's smile rises to its lips.

Perfection of human consciousness is not, however, to be sought in a regression to infancy. Mahakasyapa's smile is a smile of nakedness, but a nakedness in the three realms—dream, dreamless sleep, and waking perception. An infant's smile is in his world, together with dreamer's and ego's. An infant's smile, its silent diffusion of laughter, is a gift wrought by the matrix of life. Life when whole manifests with a smile Mahakasyapa's is an earning, a fruit of personal effort and self-initiated labors. It is an outcome—if *outcome* is the right word—of a commitment, tested by ordeal, and ripened by practice. It is, perhaps, an astonishment that the prize sought is already possessed and that, therefore, there is nothing, there has been nothing, and there will never be anything, to seek. The wholeness from which Mahakasyapa's smile floats upward stands both before and within him. His vision concludes he is not separate from the world that smiles, with the Buddha, at him. His smile is, therefore, not his but an upturned lip of reality, welcoming the momentary guest of existence.

A thinker may ask, "What is the survival value of the laugh?" A pragmatic approach, typical of a metropolis, looks for an argument with laughter, one that imbues it with a power of adaptability for *Homo sapiens.* One could speak of how the laugh unflexes us and leaves us more flexible. We may benefit from its repeated pointing to our rigidity, fixed attitudes, and unbending precepts. It also awakens perception and refreshes a fatigued outlook. Suffering and weight of age are thereby reduced. The laugh is tonic to a good life.

What a ludicrous way of speaking! As if the laugh were one function in a bevy of functions that made up life. Such logic is frivolous insofar as it stops short of asking what life is for—asking, that is, the question of the laugh. No doubt a laugher lives longer and is prone to fewer diseases (unless ticklishness is one). But the avatar of ease itself, the laugh, releases one from life to Life. True, the one who laughs is able to accomplish many great things, but the high-spirited guardian comes with a different aim. Ease, harmony of the moment, is a posture of relation to one's situation. Absent from ease is restless expectation, a mind wagging the rest of the person. Being related, ease is able to respond to Need (Love's mother) without tension or retention. Ease always faces this relation, forever background to all human endeavor.

Ease, therefore, looks in a reverse direction. We whose concern is to leave a mark upon the world confront the outer with expressive energy. Ease is mind of the other, inward current. Together, ease and action complete a circulation. The breath also has a similar cycle, an incoming, energizing phase and an outgoing, enervating phase. The cycle is delicate and subject to disharmonizing forces from body, mind, and feelings. When imbalanced from, say, shallow, fearful breathing, circulation is incomplete, leading to rapid exhaustion of energies. The same is true of the deeper cycle of action and ease, ease and action. Yet our habit is to hold our breath just before an onset of acting. Not to breathe ease in or to hyperventilate action out produces a cataclysmic fatigue. One quickly ceases to be master of an undertaking and instead is "done to." Aim is lost, and act becomes simply an outcome of influences, not of our design or intention, impinging on the agent.

Reasons of health and exigence dictate the way of balance. "The world is ruled by letting things take their course" says the same thing; "it cannot be ruled by interfering" (Lao Tzu). To cease holding the breath is to allow circulation to reestablish itself. Similarly, to cease spending oneself in action is to let ease renew a sense of self. Both are nurturing of being. Yet tension, once habitual, becomes invisible. Breath held, act conspired after—the grasp must be forcibly broken. Just here enters the spirit of laughter. Try to hold on now! Fingers tickle in the grasp, muscles melt, and the breath, natural and unstrained, once again fills the lungs. Aahh. Good.

Shiprock, New Mexico, 1997 © Arno Rafael Minkkinen. Courtesy Robert Klein Gallery, Boston

Breath, Air, Voice

Andrea Olsen

"How do we get from here to there?" he whispered in my ear. "I want to go there and there and there." He was pointing at the pond, path, and road beyond the fence, beyond the locked gate where we stand. "Sometimes the door is open; sometimes it's closed," I whispered back. "Right now it's closed." And so he waited, until he could find his own way across quiet waters.

Change is a process. "Do you recognize yourself at this age?" a friend asked. And I didn't. But I began to notice the details. Hands learn patterns and gestures throughout a lifetime. My father could no longer open any package, fit a key into its lock. "I just don't want to die fast," he said.

In my mind, a boat floats away from the shore. I am sitting on a log, listening to the waves. He raises his hand as a sign, and I know. Now that he has gone, I make dances and plant daffodils in the garden. The air already holds his footprint as memory, marking a boundary between what was and what is to be.

Each of us has passages to make, thresholds to cross, our own rituals of change. Cycles interweave, connecting bone to soil, tears to water, and breath to the air moving through us as we speak.

Perhaps the most sensuous experience is the feel of our own breath—that purr of life happening deep within the body. Breath links outside with inside on each inhale, and inside with outside on each exhale. Air moves in our bodies like an animal, filling our lungs, feeding our cells, activating our voices, compressing our skin, and stimulating our moods. Air connects our bodies to all other aspects of earth. The light waves, sound waves, chemicals, and felt sensations that give the words on this page meaning must travel through air.

We are embedded in the sky, although we often think of it as "out there." Earth's forty-fifty-mile-deep atmosphere supports life as we know it. Yet many Americans are oxygen starved, holding their breath and tensing their muscles. There are at least two phases to the breathing process: lung breathing and cellular breathing. On each in breath the diaphragm descends toward the pelvis, creating a vacuum that draws in oxygen. The oxygen must be absorbed into tiny capillaries in the lungs, transported to the heart, and pumped throughout the body so that every cell breathes. Bone cells breathe, muscle cells breathe, nerve cells breathe, absorbing oxygen and releasing waste materials moment by moment.

If we constrict our bodies, breathing is impaired. Postural habits and tight clothing that interfere with breathing are dangerous to our health. As recently as the 1870s, women wore corsets, cinching the waist into an ideal sixteen-inch circumference. This greatly impaired the capacity of the diaphragm to take in a full breath, resulting in a lack of oxygen to the brain and atrophy of internal organs. Fainting spells and "weak constitutions" were common characteristics of fashionable women of the period. Then the bicycle changed dress codes, dancer Isadora Duncan tossed away her corset and shoes, and Delsartean techniques of breath and movement (along with the brassy acts of vaudeville) challenged women to take deeper breaths, improving their health around the turn of the twentieth century.

Yet, even today, many people restrict their intake of air. In my women's writing group last year, comprising active professional women in the prime of their careers, we found breathing to be a challenge. Our group had worked for several years, performing a theater piece composed of our writings. We thought we might enliven our work by exploring voice more deeply. Our voice teacher, Normi Noel, had worked with NYU psychologist Carol Gilligan on a highly publicized research project exploring why young girls lose their authentic voices at puberty. She reminded us that many of us choose a voice in that emotionally charged period of adolescence and use it well into adulthood, long after we have outgrown those particular emotional parameters.

To explore the dialogue between breath and voice more thoroughly, our group began by telling stories, each speaking briefly about one pleasure and one difficulty—"the rose and the thorn"—in our lives. The conversation was revealing. Eyes stared at the floor. People gestured uncomfortably as they spoke. Then Normi suggested we try again, looking at each other and breathing before, during, and after our words. This simple shift required that we felt what we said and that we allowed the listener to respond. In the speaking process, our deepest self—our silent, breathing, private self—is made audible. The diaphragm determines volume, the vocal folds create the vibrations, cavities influence resonance, and the mind determines intention. As sound vibrates inner tissues of the body, we feel what we are saying, touched by the felt sensations of emotion.

Vibrations resonating in various cavities of the body add different color to sounds. Lower sensual and heart-stimulating pitches, in particular, vibrate in the large, resonating cavities of the pelvis and thorax, while higher pitches resonate in the head sinuses. Voice can be separated from feeling by restricting vibrations from moving down into the body, dissociating our feelings from what we say. Repressed emotions implode back into the body, contributing to ulcers and indigestion. Taking time to feel what we say reduces the necessity to speak, like taking time to taste reduces hunger. In many ways, embodied voice supports the underpinnings of rational thought—ethical choice made as an emotional response to various situations.

Sound evokes surprising effects in the body. When I do breath and voice exercises with my college students, giggles, tears, or a sudden chill in hands and feet stimulate departure. I know this pattern well. For many years, I would leave voice workshops. I would begin the exercises feeling the gentle flow of breath over lips, adding audible sound for a hint of vibration, growing more adventurous as a plethora of tones from various other people filled the room. For a long time, there was always a moment when the rawness of unmediated sound became overwhelming. I still can't find language to describe the sensation. It is a primitive power. I would slip out the door, assuring myself that next time I would stay.

Sometimes when we breathe deeply, we encounter grief—or its close relative, anger. A few years ago, a Tibetan student, raised in Dharmsala, India, began her senior thesis, "Youth in the Tibetan Diaspora." Midsemester in my "Introduction to Dance" class, she stamped her foot and exclaimed, "Whenever I breathe fully, I feel so angry." In creative work, anger is often a gateway emotion, an initial reaction leading to more diverse terrain. We develop gateway emotions to protect us from feelings that are overwhelming, dangerous, or beyond our integrative capabilities. Considering all she had been through, all her country had suffered, all the topics her thesis immersed her in each day, anger was a justified reaction. Working with breath, voice, and movement helps us learn how to access these emotions and channel them toward creative expression.

We can also look at muscular habits to notice breathing patterns. Some people are reverse breathers, lifting the shoulders up and flattening the belly to breathe in, so that the diaphragm barely moves. Some people "over breathe," working too hard to control the process on both the in breath and out breath, and some of us are shallow breathers, barely moving the bones and organs. When we don't want to give in or to be seen, we hold our breath. Although there is no "right" way to breathe, it is useful to develop an attitude of open exchange with the environment. Tension restricts movement and limits sensation, masking the sensitive interplay of breath.

Efficient breathing and vocalizing is based on feeling at ease. The autonomic nervous system governs breathing and visceral response, providing heightened clarity in normal functioning and instantaneous action in cases of emotional stress and threat. When you walk into a room and feel as if you can't breathe, you are experiencing your autonomic nervous system at work. The sympathetic division is responsible for what has been called the fight, flight, freeze, or friendly survival response. When threatened or insecure, you can attack, run away, or try to disappear; you can also smile, stick out your tongue, yell, or converse politely as a protective mechanism. In other words, talking and acting friendly can be a form of autonomic protection. In contrast, the parasympathetic division encourages rest and recovery, vital to health. When we feel safe or at home, breath deepens and skin relaxes, so we take in more of the world around us. Together, these gut responses motivate our actions and reactions, establishing what feels right and wrong in moments of choice.

When I teach about the fight, flight, freeze, or friendly response, I am reminded of when I was a junior in college, leaving for a year abroad in Paris. Our group met in Chicago, and we were assigned individual hotel rooms. As I went upstairs, a man in a black suit stood close behind me in the elevator, stepping off at my floor. When I unlocked my room and entered, he put his black, shiny shoe in the door—I remember the shape in detail—so that the door wouldn't close. Then he forced his way in, locking the door behind him. What surprised me most was my response: I launched into a tirade about how it was men like *him* who made women like *me* afraid, and I continued in an irritated and commanding tone. Before I had finished, he left. In the moment of decision, my autonomic nervous system registered that I couldn't fight because he was stronger, and I couldn't flee because the door was locked. It was useless to freeze, so I talked. This "friendly" response—our capacity to breathe, smile, make facial gestures, or otherwise verbally persuade—is a survival mechanism that many of us use throughout the day.

At a young age, we internalize habits of breathing that can be useful or harmful to our lives. Breathing patterns can reflect the rhythm of the mother's breath in the womb, the birth process itself, or later impressionable events in life that become fixed in nerve pathways. As a baby, I would hold my breath, turn blue, and faint. This occurred from six months to two years of age, and once—as the family story goes—necessitated a shot of adrenaline in my heart to induce revival. Midway through my first college faculty position, I developed a condition where I couldn't breathe. I would go to a movie, see an image of a woman drowning or an abandoned child crying for his mother, and not be able to catch my breath. I was given mild muscle relaxants and told to rest. Then I attended an experiential anatomy workshop. I was told that there is a point in the breathing process where you have to give up control. It is the point of release that allows exhalation. It involves trust that the next breath will come.

A month later I observed movement educator Caryn McHose teaching a class. She had the participants curve into the fetal position and place their hands on their lower backs, feeling the movement of muscles under their palms as they breathed. She called it the "breathing spot" and said that the whole spine should move when we breathe—all the way to the coccyx, or what's also known as our ancient tail. I saw the fluidity of her body but was unable to feel it within my own structure. The following week, I awoke to the sensation that my spine was moving. It was undulating gently with my breathing, like a river flowing inside me.

Deepening our relationship to breath is a lifelong process. From first breath to last, air mediates the dialogue between self and the world, becoming a symbol of life. A healthy baby bonds with air on the first breath, and with the mother by nourishment on the first suckling. Humans require connection to air, earth, and nourishment for survival, as well as touch, movement, and community. When my mother suffered a dangerous stroke, I could hardly breathe. I didn't want to take in the news, to absorb it in my body cell by cell. I felt the ways I had been supported by her breath all these years—the enormity of the umbilical connection. I touched her hand, caressed her cheek. At some point, I knew her breath would stop and somewhere else a new breath would begin, reminding me that the only constant is change.

Rhythmic cycles are basic to both body and earth. Rhythm occurs on various timescales, from the repetitive sound of a footfall-heightening awareness of breath, to the seemingly isolated eruptions of volcanoes creating the chemical balance of our atmosphere over billions of years. Rhythm is also a way of establishing community. When we move and breathe together in rhythmic interaction, we develop a relationship. When we are aware of the rhythms of the seasons, epochs, and eons, we feel ourselves as participants in an expansive choreography.

From an environmental perspective, breath and air are one. Humans are subject to the same physical laws as all other components of earth. A lung branches the way a tree branches, because they are governed by the same forces. The body is home, our first environment, the medium through which all experience is channeled. To inhabit place with integrity, we inhabit ourselves cell by cell, recognizing our role in larger systems. Humans live somewhere between the cool perimeters of sky and the fiery core of earth. Once we envision ourselves inside a living system through the intake of air, we recognize our participation in the world. It is no longer a question of paying attention to the earth and the sky; it is a process of living our relationships with consciousness.

Breath is inextricably linked with rational thought. We must be cautious, in this fast-paced world, that we don't hold our breath, dampen emotional responsiveness, and lessen our sense of responsibility. When we cease feeling, we stop caring. The language of the body helps us to respond to each moment. As we feel sensations, breathing in and breathing out, we make informed choices, allowing our bodies to guide us responsibly in our interactions with the world.

Acolyte

Steve Miles

Walking the river,
        a sliver of stink
                slips inside, first taken
for dead algae
        chuffing its gas
                as the river dries.
Then, a perineal musk
        says *skunk!*
                Sulfuric methane,
sultry & steroidal,
        a rectal simmer
                that can pry bark
from trees, it swells
        like welts
                under my eyes—
Irrevocable.
        My heart downshifts
                through farmyard junk,
the fetid funk
        makes a whitewall tire
                jump off a stump.
Memory knots
        the intestine of fear
                I pilgrimage toward,
silage of a sunken god,
        garden sod,
                a pleurisy of bees
in the fallow trunk,
        turning the yeasty
                compost in with my dad
still drunk, breath
        hunkered under
                a spade where I first
saw some things
        shrunk dead never
                Lazarus back, they only
grow black mold bursting

under thumbprints, ashes
　　　　to ashes, solids to gases—
the spirit needs a little
　　　something to cling to—
　　　　　a hex-noose of ether
cloying the tonguetip,
　　　　pheromone, mercaptan,
　　　　　voodoo rubato,
flammable fumes
　　　mummify the river,
　　　　　swim open gates,
tumble through locks
　　　as all souls
　　　　　un-spelunk
& ride the inky
　　　aldehyde
　　　　　of skunk.

Luberon with Mist, Cherry Blossoms, 1999 © Crystal Woodward

A Breath of Memory

paulo da costa

Florindo Ramos loved trees since one turbulent winter day when, as a limber boy, he had been hopping from stone to stone on the edge of the River Caima and suddenly slipped, plunging into the careening current. "I'd have drowned if it hadn't been for my Ginkgo, growing by the lip of the river," Florindo told the children who gathered around him after school, listening to his tales. Lovingly, Florindo stroked the glossy green leaves of the saplings, growing under the canopy of mother Ginkgo. "I helplessly flapped my arms, rapidly sinking. But my Ginkgo leaned almost out of the ground," he pointed to her exposed roots which resembled legs, "and she swept her arms into the waters to save me." Florindo tenderly kissed the reddish tree bark while leaves stirred shyly above children's heads.

Following that miraculous winter day, each night after his father retired to bed to say the rosary and dream of sacred mysteries, Florindo climbed out of his window to the Ginkgo. He slept curled around the tree's feet as the River Caima flowed and gurgled at his toes. His palm on the trunk, he felt the Ginkgo's pulsating sap pulling nourishment from deep inside the earth.

Awakened in the morning by the yellow warblers' perfect songs, Florindo rose and, stretching his arms, brought the Ginkgo's flowering buds to his nostrils, filling his lungs with the pungent odor. He yawned and rubbed her leaves between his fingers to freshen his skin. The tree breathed out, Florindo breathed in. He breathed out, the Ginkgo breathed in. A conversation that enriched and nourished their lives.

Ti Clemente and many of the village people resented Florindo Ramos's apparently useless life. While Ti Clemente headed for the fields with the sunrise and toiled until the last crumbs of light fell from the sky, Florindo sat under the shade of his Ginkgo surveying the world.

"Should put his chit-chat rigmarole to good use behind the ploughing bulls, convincing the beasts to budge," Ti Clemente snorted under his breath, leading a team of bulls to the field with the prick of his driving pole.

"But he's in love with his maidenhair tree," excused Ti Clarissa, walking behind Ti Clemente. From time to time, she secretly brought corn bread to Florindo. "He's happier than any of us. See the attention he pays her," she pointed at Florindo brushing the Ginkgo's trunk with a corn broom.

Florindo did not respond to the swing of Ti Clemente's axe-like tongue. He finished grooming his Ginkgo and then sought solace from the day's heat beneath his tree while her fan-shaped leaves gently swayed and stirred a breeze, cooling the air.

Nose to the grass, sprawled under the Ginkgo, Florindo awoke one afternoon contemplating no particular thought. He watched a bumblebee collecting pollen, when suddenly the crystalline words of his deceased grandmother flooded his mind. Florindo slowly remembered the days they had strolled together across cornfields, along the River Caima, picking bouquets of wild violets from the southern slopes. "Come closer. Bathe in their perfume," his grandmother had whispered, bringing the bouquet closer to his nostrils. The delicate petals had tickled his nose.

Florindo gradually realized that the same air that had dwelled in the field and in his grandmother's lungs now stirred in his chest and brought him pictures of the past. He smiled and clacked his tongue.

"Memory is breath," he exclaimed aloud and placed a blade of grass between his thumbs, blowing a long shrill. Florindo expanded his lungs with air, thinking how the moment lungs emptied, life ceased. People stopped remembering—remembering who they were, remembering the habitual traps of past deeds. Memory disappeared on the wings of the last breath, and there was nothing, only the vacuum of death.

Years later, in a city hospital, visiting his father, Florindo faced a faint line that filled his father with forgetfulness, edging him toward death. He ran his fingers along the tube entering his father's nose, examined the bottles pumping air into his body to keep him alive, and realized his father was on a mistaken path to regain life.

"You're feeding him lifeless, bottled air," he complained to the doctors. "My father must return home. He must lay among his family, sharing the air we breathe to remind him of who he is, where he comes from!"

The doctor stared at Florindo's reddish face and the scattered leaves nesting in his locks. The smell of fresh grass permeated the air.

"I understand how disconcerning it must feel," the doctor said, patting Florindo's shoulder, offering him a glass of water and a fluorescent pill. But Florindo's eyes were mesmerized by the large poster of a human lung hanging on the hospital wall. He walked closer to the picture and with his nail smoothed out the paper's creases.

In the picture, the lungs resembled the tips of tree branches. "It's here the air whispers in and out," Florindo said excitedly to the doctors. It was as if people carried a miniature forest inside their chests. "Hmmm . . . the alveoli," nodded the doctor. He burst out laughing and Florindo ran, short-winded from the stagnant air of the hospital.

That evening, the Ginkgo nestled Florindo in her branches, rocking him back and forth as he cried himself to sleep. In Florindo's dreams, the Ginkgo whispered that he must grind her leaves to dust, mix them in water, and offer it as a drink to his father. "It'll open the air flow to his heart. His memory will improve," the Ginkgo insisted, in the assured voice of the oldest lineage of trees on earth, survivors of countless trials.

With no sign of a breeze and as if nudged, Florindo awoke from his dream stretched out on the ground. The Ginkgo leaves rattled frenetically. Florindo hurried. He collected a handful of her leaves and ground them with a stone on a flat smooth rock. Then he ran the distance back to the city and slipped into his father's hospital room with the Ginkgo paste in his pouch. He opened the window in the room and spent the remainder of the night moistening his father's lips with the concoction.

In the morning, when his father regained consciousness, the doctors sent him home, "Spontaneous remission," they concluded, scribbling in their medical charts.

Dark clouds gathered over Vale D'Agua Amargurada when Florindo returned with his father. The pines and oaks brandished their branches. A wind lifted clouds of dust. A cold hiss swept across the fields. In the schoolyard, a group of children played, galloping stick horses, sword fighting in a battle of knights. Leaves rattled incessantly like grinding teeth as lightning tore the sky to rags. A tree creaked and crashed down on the sawmill. Drops began to fall.

Florindo ran to his tree. Tearlike sap dripped from a broken limb on the Ginkgo, while at her feet, her crop of saplings had been ripped up and torn. "I see you tried to push away the thoughtless plunderers," Florindo whispered, placing his face on her moist trunk, running his hand down her back. Tearing his shirt in strips, he bandaged her limb and carefully improvised a sling.

Florindo felt the earth tremble as the pitchfork of lightning bolts sank into the ground. The River Caima rushed, sparked by the lightning that traveled through its vein. The hairs on Florindo's neck stood up. Hurriedly, he gathered armfuls of leaves and buried the saplings. Then he walked to the school yard and collected the children's abandoned swords. Under the Ginkgo, he thrust them through the leaves and into the ground, where they rose like beautiful crosses. "You must be buried in the place they were born," Florindo solemnly said and played mournful songs on his harmonica until the rain stopped.

After the deluge, the children tentatively stopped at Florindo's Ginkgo on their way home from school, books tucked under their arms.

"Florindo, Professor Manecas says we should be doing our homework rather than wasting our time talking to you," said Bonifácio Careta, who was the most skeptical of the bunch and always stood, afraid to dirty his yellow knickers.

Florindo sighed to the Ginkgo. Then he replied.

"Come closer." He motioned to everyone. "I'll tell you a secret." The children huddled around him like fiddlehead ferns. "A tree grows branches in many different directions," Florindo said, lifting his thumb upward to the Ginkgo. "If you follow the aim of each finger, you'll see many different things, things you otherwise might not notice."

The children looked upward, each finding a branch to follow. "There's a pigeon's nest on the church steeple!" "And a rabbit's den across the river!" "A horse cloud!"

Florindo nodded and proceeded.

"Professor Manecas teaches you that knowledge is in books. But remember, books are made of trees." The children gaped at their books, then at the Ginkgo. Florindo continued. "Maybe those who write books think they concoct things in their minds, but they are fools. The knowledge was already there, waiting to be remembered and inked."

Florindo wetted his index finger and flipped open a school book. He rubbed a single book leaf between his fingers. The paper crinkled. "That's very thin knowledge compared to the trunk of a tree," Florindo said, and then made the ginkgo leaf dangling from a corner of his mouth twirl along his smile with the movement of his lips. "Sit under a tree long enough, and you will also know things," Florindo concluded and took up his harmonica as a light drizzle began.

The village children eagerly ran to visit Florindo's after school, and they listened to the wondrous stories that shed light on the workings of the world. His stories were as sweet as the wild honey that dripped from the beehive in the Ginkgo's upper branches and as tart as the bite of a ripe olive.

"Florindo, don't put monkey ideas in their heads," Ti Clemente yelled from afar, half-chuckling, walking back from the fields with a bundle of grass on his head, destined for the corrals.

Florindo's ideas had been called "monkey ideas" since he told the children that people descended from trees. Florindo walked to the lumber camp at the edge of the village and gathered the children around a stump that was as wide as a table-top. He placed his open palm on the stump, next to the rings on the wood, and showed the children that people and trees were cousins. The awed children released a collective whistle. The resemblance of the contour lines on the dead tree and their own fingertips was undeniable. "The trees left their imprint on our hands from the time we dangled from their branches," said Florindo—a story that fell from grace the moment it reached Padre Lucas's ears during catechism and required much holy water to dilute its heretic ideas.

The children continued to flock to Florindo's presence. Florindo not only told stories, but also gave his full attention to their own. He listened to shy Alzira as she remembered being a bullfighter, describing in minute detail her foot dance in front of the bull, the colorful banderillas in her hands, the inciting sound of the trumpet.

Florindo believed that traces of past lives lingered in children's bodies. But as they aged, thoughts of the present left no room for the past.

Senhor Mário Mateus peered from his window and waited for the children to be called home for supper. Then, in the company of his faithful umbrella, ensuring no one noticed the detour from his daily stroll in the hills, he stopped by Florindo and his Ginkgo tree.

"Florindo, let's say you knew of someone in need of advice on how to clear the nightmare of a haunted mansion. What would you suggest?" Senhor Mário Mateus shuffled his feet, staring at the sky.

Believing every house carried remnants of ancestral presence, Florindo did not believe in haunted houses.

"A house overburdened with the breath of the dead is lopsided," Florindo told Senhor Mário, and pointed at his crooked mansion's roof, sagging under the load of the years. "To balance, open the mansion's glassy eyes to a sunny day and invite children's laughter to share your dwelling. Children's play, as they run to and fro, keeps the air moving."

"That's a rather complicated solution. Do you have anything simpler?" Senhor Mário retorted. Florindo ignored his comment and continued.

"A well-aired house restores balance."

Senhor Mário Mateus scratched the ground with the tip of his umbrella and muttered indiscernible sounds before staring at the sky again.

"And what about ridding ancestral ghosts in just one bedroom then?" Senhor Mário Mateus said as he straightened the rose in his lapel.

"Ghosts want to be heard," Florindo said. "Ancestors' thoughts—trapped in the wall tapestries or clinging to the dust on yellowed photographs—are as essential as bread and water to a joyous home." Florindo paused and noticed Senhor Mário's hands, his knuckles white from clenching his umbrella. "Sometimes the dead require the hands of the living to repair wrongs they rendered in their lives. Once we listen and cooperate with their quest for atonement, the dead will leave us to breathe in peace."

"I'd better be going before the sun dies on me," Senhor Mário Mateus said as he hurried away and into the hills.

Perched on the Ginkgo's highest branches, Florindo stared at the hillside across Vale D'Agua Amargurada, the hill continuously carved to terraces for village crops.

Florindo leaned on the trunk of the Ginkgo, cupped his ear against the bark, and listened to the moan of forests, near and distant.

"Dear Ginkgo, what can we do to appease the trees? Ti Clemente and the others don't listen to us!" He stared at the decimated hillside.

Florindo believed the world's knowledge entered trees through their leaves and needles and the irrecoverable story of the world was buried in their roots. The trees stored their thoughts in their roots and if turned into stumps, they became unable to exhale their memory, unable to release their stories. Scarlet fallen leaves were cut-off tongues scattered on the ground. Lives cut short, denied of old age and denied the slow farewell years of passing their accumulated wisdom onto new saplings.

As the hillside forest fell, Florindo noticed decay in the village mood. Ti Anastácio, who had never swatted a fly, began kicking his long-time canine companion, Peyto, curled around his feet, anytime the nearly deaf creature did not obey his orders to move. Felismina Alves, the village *curandeira* who for decades had burrowed in the reclusive heart of that south slope, could not cure the simplest cold. She appeared to have forgotten her spells and the whereabouts of her sacred ingredients.

Ti Clemente broke agreement after agreement, alleging he did not remember shaking hands on it. People grew obsessively ensnared in the mudslides of the past, forgetting the essence that held fast the soil and grounded their soul in the world.

At the end of a village day, the sky perspired in a blush of burgundy, and boys clustered under the Ginkgo tree. They wrung their hands, requesting advice to quiet the nagging awkwardness of their enamored hearts.

"Will you make up a lovely poem I can give Rosa?" Armando asked him, a rosebud tucked in his ear.

"Love is not embedded in words," Florindo said. "To be certain of love's presence, to be certain that you aren't ensnared in the mirage of the eyes, you must kiss." Florindo kneeled to a daisy and drew a deep breath. "Kiss deep, kiss long. Invite Rosa's fervent breath into your lungs. Keep your spirit door open. Your tongues, guardians of spirit, will be silent in their dance. They will commune. Unimaginable feelings will stir and surface. Your histories will mingle. You'll be of each other." Florindo paused and nibbled on the white petals. "You'll know truth, each other's innermost secrets. Be delicate with these treasures. No matter how many hills may rise between you and Rosa, scent and taste will link you forever."

A breath of wind dropped a handful of blonde leaves. They fell in Florindo's lap. He blushed. Autumn had arrived to undress the Ginkgo. Florindo sat shyly next to his beloved tree.

"Florindo, if you are so in love, why don't you wed your Ginkgo?" Bonifácio Careta asked. He was perpetually eager to play matchmaker and never passed up an opportunity to suggest a wedding.

Florindo winked at the Ginkgo. He closed his eyes, placed his palm on her trunk.

"Now, that's a sweet idea," Florindo said.

On Sunday, the children hung garlands of violets from the branches. In the ceremony that Padre Lucas refused to bless, there were no words spoken. Felismina Alves tattooed the Ginkgo Biloba's initials on Florindo's arm, and he carved his on the tree with his thumbnail. In their beaks, birds from the riverside bushes ferried blackberries and showered the guests. The children lay on their backs, mouths open, tongues ready to snatch the cascading berries.

The Ginkgo's leaves rattled like castanets, and Florindo, clicking his heels, swung around her trunk dancing his happiness while Prudêncio Casmurro played the concertina.

The villagers pretended the idea was insane. But sitting on the opposite side of the River Caima, fishing or weaving orange osier twigs into baskets, they followed the merry events.

"I guess you'll move in together onto a tree house," Ti Clemente yelled across. The women shooed him, clapping to the concertina. In the din of merriment, his words never crossed the river.

December cold had settled in the valley. Women were busy in their kitchens, baking the fruitcakes and soaking the salted cod for Christmas Eve. Ti Clemente and bands of villagers, axes on their shoulders, marched into the woods in search of the perfect pine tree. Florindo followed them in silence and lit candle lanterns on the stumps of the fallen trees. At night, next to his Ginkgo, he sang his prayers above the drifting hymns of the congregated at church celebrating the birth of their Creator.

Florindo celebrated Christmas by hanging olive oil lanterns from the fingertips of his Ginkgo and spreading cotton candy over her branches. One by one, after midnight mass, the children sneaked out of their houses and joined Florindo, perched on the Ginkgo, singing and licking the cotton candy. The children in their colorful garments dangled from branches, their scarves swaying in the air, resembling living tree decorations.

A bonfire, from the Ginkgo's fallen leaves and branches, roasted Ginkgo seeds. The white kernels Florindo regarded as a delicacy.

That evening while the adults slept, Florindo and the children spent the translucent night planting seeds on the periphery of the village. They planted and sang, "Dance wind dance, waltz a seed, you must believe, trees will live."

Year after year, unnoticed by the villagers, trees regenerated around the village, forming a solid protective shield of greenery.

Village men gathered around the transistor radio at Ti Anastácio's *adega* listening to the evening news while Florindo dangled from the crown of his Ginkgo listening to the wind. He sat motionless, eyes closed, letting his belly, his fingers, his toes, fill with air.

Florindo felt the stirrings in the air as the tingle of adverse news gained momentum on a wind. His fingers pulsed. Florindo listened. The villagers turned up the radio, turning a deaf ear to the subtle stirring in their spirits.

Florindo could see the skies darkening to the west. The trees swayed in the wind, rattling their leaves as clouds of dust hovered in the air.

The children curled up beneath the great Ginkgo. To ease the children's fear, Florindo assured that the enraged winds would not touch them. "They would not harm a cluster of their own," Florindo whispered gravely, pointing to the shield of trees around the valley.

Through the night, the Ginkgo trembled and creaked against the iron wind ploughing the land, sending cattle spinning in air like pebbles. The Ginkgo stood her ground, and Florindo and the children slept curled around the tree's feet.

Florindo awoke to the hymns of the church congregation floating across the river.

The hurricane that had swept the countryside had miraculously left the village relatively unscathed. The villagers marveled at their divine fortune. They promised to pray more fervently and to be faithful to the Ten Commandments. Then, in praise to God for sparing their lives and land, they swung their axes over their shoulders and marched out to fell the oldest and strongest trees to erect a chapel in honor of the Lord of Heavenly Winds.

Tuula Närhinen, *Anemographic*. Drawing made by wind and birch, part 2

Aria

C. L. Rawlins

> *Earth's breath becomes my breath, this is how I shall live.*
> —The Blessing Way (Navajo)

I want to leave the snowmobiles in town. Likewise, the damned four-wheelers. Noise. Stink. Let's leave it behind. John is skeptical. "Remember the last trip," I argue, "dragging the stupid machines out of thawed snow and bogholes, breaking through the ice? Remember how we did powerslides down that gully off Big Flat-Top? How I had to dismantle the log fence with a rock? Remember carrying the four-wheelers across Dutch Joe Meadows?" I stomp and gesture.

It's early May, after a winter of thin snowpack. It's been warm and the snowline has receded. We can hike up to the continuous snow, then ski. John isn't convinced. The packs will be too heavy. We may have to walk in a long way through mud. I bully as much as persuade. It'll be quieter. Easier. More aesthetic. More fun. John's not a motorhead, but neither does he want to plod through the lower (i.e., boring) country on foot. He tucks his chin into his collarbone and sets his sunburnt jaw.

"*¡Madrecita de Dios!*" I squawk in frustration. "Okay. *I'll* carry the beer." At that, he shrugs and finally agrees. He just finished off his regular work week with a fiesta in Jackson and looks raffish. We leave the four-wheelers on their trailer in the Forest Service yard and drive south, truck tires whining on the pavement, then climb the desert apron of the range on dirt roads, following last year's ruts, newly emerged from the snow.

John sleeps as the day dawns. Across the broad valley of the Green River, I can see the crest of the Wyoming Range fifty or so miles away. Up north, Gannett Peak, Wyoming's highest, catches the first glow on summit snowfields. From high in the Wind Rivers, on clear, cold winter days, it is easy to see the Grand Teton, Wyoming's second highest peak, almost a hundred miles northwest. From high points, looking southwest, I might also see the snow-covered mass of the Uinta Range, about a hundred and fifty miles off, stretching like a low, blue cloud along the horizon. This is about as far as I can see in perfect conditions, dry, clear air, no haze or dust. Beyond this, the atmosphere itself scatters the light, and the contrast by which we distinguish object from background is lost as the earth curves away.

We drive into sunrise, the big peaks to the east like blue cutouts under a clear sky, their long slopes cold in early shadow. Later it will be warm and the snow will melt. We want to make it in to the cabin—probably about nine miles on foot—before it does. Otherwise, we may have to posthole or even bivouac if our skis fail to support us with these monster packs.

South and west of the range, the land lies in long, depositional planes, tawny with short grass, spotted with sage or pocketed with aspen and juniper where bluffs give shelter from the wind. The land stretches the eye, the horizon at least a day's walk in any direction, or two, or maybe a week's walk to the west and south. This country opens to the southwest, toward the prevailing wind. Where the wind is strong, over ridges and in saddles, the vegetation hugs the earth. Everything from bedrock up—soils, plants, animals, weather—depends on how the wind blows. It's the air's country more than ours.

We think of air as being the same as space—open space. This is easy to do in the West, where the air is most often dry and clear; out here, the sunset is fifty miles west. In humid regions, the air has more presence, coloring with changes in the light; there, sunset happens all around you. You can almost swallow mouthfuls of it. In cities, the air has a kind of malevolence, the medium for oppressive smells and tastes, but even there we think of the air as a kind of void, a region where things as we know them don't exist except in transience: the silver knife of a jet, a baseball rising from the bat, a flock of pigeons, wheeling.

Yet the earth's air is a vital medium, vital in more than one sense of the word. It is intimately connected with organic life, in fact partly the result of it. To that life, our life, it is essential. It sets life's boundaries by changes in its temperature, movement, and composition. And it is an organ of the biosphere, as crucial to the continuance of life on earth as lungs or heart to a human body.

In comparison with the mass of the earth, the mass of the atmosphere is not large, nor is it of great extent. The zone habitable by humans extends from near sea level to about fifteen thousand feet, just under three miles. Above this, the pressure of the oxygen is not sufficient to permeate our blood. Humans can adapt to work or climb above this elevation, but permanent settlement is impossible. Higher still, say above twenty thousand feet, the, body will break down. Mountaineers call it the Death Zone. Among the great distances of planet and surrounding space, we have livable atmosphere only three miles deep. Compared to the earth's four thousand-mile radius, this is like the single outer layer of an onion's skin. More closely, it compares to the thickness of the outermost layer of cells in your skin.

The biosphere—the zone supporting life—is somewhat larger, from the ocean deeps to the highest point in the atmosphere at which floating spores can survive. The zone affected by life is still deeper, from the strata of carbonate rocks, like limestone, deposited in the crust, to the stratosphere where earth's distinctive, biologically oxygenated gas mixture burns up meteorites rather neatly.

It's doing a good job. We don't see a single meteorite as we bounce around Big Sandy Ranch. Years ago the ranch belonged to the Leckie family. When I got here, it was owned by a handsome ski-hound who spent winters in Jackson and summers down here. Now, it's occupied territory, snapped up by a man from New York City, but unoccupied at this season. To the rich indoorsman, this place is habitable for about three weeks out of the year—the span between mosquitoes and the first snow. Turning north, we labor the green truck, battered as most Forest Service trucks are, up switchbacks, climbing out of the valley. I roll down the window and the air comes in cold. I take a breath, then put it back.

Without life, our planet's atmosphere would be like those of Venus and Mars, more than 95 percent carbon dioxide, around 2 percent nitrogen, with no free oxygen at all. The mean surface temperature of earth would be more than twice the boiling point of water. Instead, we have about .03 percent carbon dioxide, 79 percent nitrogen and 21 percent oxygen, our mean surface temperature is about 13°C, or 55°F.

There are purely physical models that attempt to explain the development of our atmosphere, but they all grind to a halt, with oxygen and nitrogen locked in compounds, hydrogen escaped, no free water, no available energy. The major physical difference between the earth and Venus or Mars is temperature: distance from the sun. Yet, by this measure, our planet should be much hotter than it is. The habitability of our world owes to our unique atmosphere, which is biogenic: created by living organisms.

What does this mean? Simply that our present atmosphere is the sum of millions, perhaps billions of years of life and death. The first life is thought to have been anaerobic, unable to live in the presence of free oxygen, like the bacteria that still live in the mud of swamps. Anaerobes also live inside us: in our guts, where they produce methane, the same natural gas we drill for and burn.

Obviously, I've been reading about the air. I try, between bounces, to tell John about it. He knows quite a bit, it turns out, but he's not in the mood for science.

"I had a lot of that shit in classes," he says. "I don't know why I remember it. They told me I could get a real job. Lies." He scowls and burrows into a pile of coats.

We pull out on top of a hill, the ruts getting more serious, the snowbanks deeper. We creep through a patch of thick lodge poles, avoiding a wallow-pit where someone got stuck. We cross an open meadow and see the wallow-pit of all wallow-pits, a Rock Springs Special. Someone got stuck once, got unstuck, and then, in defiance of good sense, roared back, trying to blast through a deep drift. The truck went sideways and then spun out, hacking its way into the snow like four chainsaws. I can see the tiremarks, shovelmarks, kneemarks, and bootprints. Cowboy boots, needletoed, lowheeled. Truckdriver models. Two pairs, about sizes 10 and 12. Chunks of aspen and lodgepole, jammed and jackstrawed. Mud sprayed out from deep gouges onto the snow. Mats of sagebrush, mashed by spinning wheels. Cigarette butts. Seven empty cans, Coors Light. They must have been stuck for at least an hour. They were here last night. The tiny sagebrush leaves are still soft, their broken edges still moist. I can smell them, a sharp green scent.

Photosynthesis, the trick that leaves do, didn't happen overnight. Before there were plants as we think of them, simpler specie—called cyanobacteria for the blue-green pigment cyanin—started using sunlight to fix carbon directly from the CO_2—rich atmosphere. The carbon-based compounds it yields are called organic matter. The photosynthesizers, a sort of chemical mirror image of the anaerobes, began giving off oxygen and other gases.

In photomicrographs, cyanobacteria look like the beaded name bracelets, that hospitals placed around the wrists of babies in mid-century. The oxygen they gave off reacted almost immediately with the rocks or atmosphere. But colonies of photosynthesizers were the first habitat for complementary organisms that evolved to produce energy by oxidation. These creatures—I don't mind claiming them as family—were the forebears of all animals, a word whose root means, "breathing one."

We park the truck, stuff our packs until we can barely lift them, and set out. Our breath forms clouds, each curse visible. Dutch Joe Meadows is a muskeg, all sponge and snow melt. We hop and dance and sidle, stumbling under the weight, trying to keep our boots dry. We reach higher ground, a sagey hill, and strike for the guard station, a two-story cabin, built by the CCC. Even the high ground is spongy. The brush grabs at our ankles. John glares at me. We hike up the hill and see that the door has been forced by snowmobile vandals. We drop our packs and go in, boots booming on the hollow floor. Maybe "vandals" is too strong a word. There's no other damage. They just wanted to see what was inside, to build a fire in the fireplace. Maybe the cabin should be left unlocked. We shoulder our packs and head on, looking over the pines at the sky. There are a few stringers of cloud, but not the kind that herald storms. We go through the back gate, a groaning old timber one, and cut through the woods, wading drifts to a low crest.

On the downslope, we strike the road, untracked since last fall, and hike up it, avoiding the deepest snow. The road is cross-fenced by drifts. The alternation of mud and snow would make tough going for a vehicle, even a loathsome little ATV. It would be like driving over a dead cow every ten feet. Most of the mud is still frozen. Occasionally, though, a patch that looks hard as rock will prove not to be, oozing up as my boot sinks. We follow the road, then veer off it into the woods, following the open ground, covered in pine needles and more solid than bare dirt. Crossing the snow-banked bridge on Dutch Joe Creek, I look down into the slow current. There is broken ice along the edge. Under my reflection, the silty bottom is blackish green with periphyton, the oozy stuff that grows on underwater surfaces. There are willow stems, chewed by beaver and weighted with rocks. I see a fine mist rising from the riffles upstream, catching light, then disappearing a few feet above the water. Most of the stream flows downhill, in its meandering channel or through the ground, but a small part of it rises into the air, flowing back into the sky.

Not just water, but water vapor was essential to the development of life on earth. It is probable that this early atmosphere included a layer of smog, the product of photochemical reaction between methane and fog. This layer would moderate temperature changes—the first greenhouse effect—and filter ultraviolet radiation, allowing other photoreactive gases like ammonia and hydrogen sulfide to accumulate below. This would have improved the conditions for life in the oceans, and the resulting bloom of cyanobacteria may have produced oxygen in such quantities that all of it did not immediately react: free oxygen, free at last.

Do I mind owing my existence to blue-green scum? Looking at my reflection filled in with stream-bottom colors, it seems natural. I can see the periphyton like a fuzzy coating on the stream's tongue, outlined by my shape, with my expression, my pair of eyes. All those little mothers working away down there in the primordial ooze. The thought makes me grin and the scum grins back.

While many complex species are extinct, the pioneer forms of life such as viruses, anaerobic bacteria, cyanobacteria, algae, diatoms and zooplankton are still common. Not only are they common, but they are also the living foundation which supports all complex plants and animals. Little mothers indeed, they are still maintaining earth's atmosphere.

Around two billion years ago, the amount of carbon dioxide declined precipitously, perhaps because of all the frantic carbon-fixing going on under the atmospheric blanket, which deposited millions of tons of carbonates on the ocean floor in microskeletal remains. Every atom of carbon thus buried freed two atoms of oxygen. This increase in oxygen would have the further effect of binding hydrogen atoms in water molecules: we all know the formula H_2O. Hydrogen atoms have so little mass that earth's gravity can't hold them: they float off into space unless contained or combined. Even water vapor is broken up into its component atoms by the low pressure and solar energy high in the atmosphere, whereupon the hydrogen dissipates into space.

With a massive sigh, John flops his skis down on the snow. I slather my bases with red klister—a wax that looks like ruby-colored honey—and clip in. Showing his first grin of the trip, John forges ahead. The snow is sleek and firm, like a cat's back. We're so happy to be on skis that we forget to watch the sky. It's still clear ahead, north and east, so the sudden rise of clouds from the west catches us by surprise.

We ski west of the beaver ponds, where there are beaver trails and fresh-cut bolts of aspen. The terrain is glacial, full of gouges and moraines, the outwash pockets filled with beaver ponds. The clouds bear down on us and pass, each one spitting out what it holds, little pellets of graupel, then big pinwheel flakes. As the surface changes our wax balks, clumping with fresh snow, and then works again.

We schuss the road down the last hill and flat-track into Big Sandy Opening noticing open ground on the south slopes and open water in the creek. John stops and groans. He has a headache. I can tell he's miserable, but he doesn't lash out or get broody. When Marty had an ache, the world was responsible. Between spates of temper, he would describe it in fine detail and ask, twenty times a day, if it could be serious. "Don't worry," I'd say for the first nineteen, "it'll go away by tomorrow." The twentieth time, I'd get impatient. "It's terminal," I'd say. "You're buzzard shit, José."

We straggle through the succession of gentle, treeless swales that flank Big Sandy Creek. The Opening, as it is called, is glacial outwash, possibly a lake at some time, about two miles long. For a place that appears flat to the eye, it has a surprising number of ups and downs. John is suffering, but brave. The skiing is good until the last mile, when the snow reaches the melting point and the Opening changes to a huge bowl of cream soup. We thrash across a blank space, herringbone the hill, and fetch up at the *Casita Perdida,* the Little Lost House.

I unlock the door and we tumble in. John mumbles something about lunch and piles into the only chair. I build a fire in the stove and pull the foodsack out, spreading the table with everything we've got. We sit and stuff, not a word passes between us. It's quiet. I like it. Without the big snowmobile sled, we couldn't bring a boom-box, to me a blessing. But John looks forlorn.

He finishes eating and clambers into the upper bunk, going to sleep in seconds. I lift the top plate and stoke the stove, then watch the flames eat the splits of pine, soaking in the heat. Oxidation. Where would we be without it?

We think of oxygen as a life-giving gas, but it is both poisonous and dangerous in high concentration. In ozone molecules, O_3, or in unstable combination with hydrogen, oxygen reacts wildly. If these were present at higher concentrations, steel would rust immediately, and the paper of this page would blossom into flame. There is a layer of concentrated ozone high in the atmosphere, where it forms in the strong sunlight. Ozone filters out ultraviolet while allowing visible light to pass. Oxygen is like the unpredictable guest at a dinner: three glasses of wine and he's delightful; four glasses turns him into a wild boar.

The proportion of oxygen is critical; with a 4–5 percent increase in oxygen concentration the entire vegetation of earth along with any exposed wood, coal, or organic soil, would erupt in a burning that would make the fire in Yellowstone seem like a single Lucifer match.

The incidence of fire is a powerful consequence of oxygen levels. On the other hand, with a 4–5 percent drop in oxygen, nothing would burn. Thus, the world we think of as ours is dependent on oxygen levels that vary within one or two percent.

In animal cells, oxidation is the major source of energy and also the largest part of the biological process we call aging. We oxidize nutrients to live and are in turn oxidized: the biological equivalent of rust. Clearly, the proportion of oxygen in our atmosphere has been maintained for a long time in order for the form of life that we know—that we are—to exist. So we not only evolved in earth's atmosphere, but in intimate connection with it. We depend on the air. The air depends on us. I smile up into the blue. It's like finding a long-lost brother.

It's too late to head for the collector. The thaw has taken over. We didn't have space in the packs for a tent—what with science junk and amenities—so we're committed to a lightweight ascent. In and out in a day. I push a few more splits into the stove and leave John to his dreams.

Outside, there are open spots under the pines on the southwest slope. Between snowbanks, the moist needles are rusty red, the earth is chocolate brown and the rocks are splashed with lichens, orange, yellow and lime-green. The storm has blown past and the sun is bright and sweet. I find a dry patch and lounge on the matted needles, smelling the rise of the sap in the lodgepoles. With a boulder as a backrest, I lay my head on smooth granite and nap in the sun. When I wake up, buttercups have unfurled around me.

We speak of time as a river. By a river, though, we can choose to walk downstream or upstream. We can swim it. If it's shallow, we can wade across. Time has no banks, unless it washes the shores of the universe. Time is the sun's heat against my skin. Time is buttercups.

The sun is seductive. I strip off a layer at a time until I'm naked, lolling on my heap of clothes. I think of women. Ideas are too pure, either snow or flame. A living body has its own, dependable heat. I recall favorite sins, matchless transgressions. I think of Bella Linda, heavy hair resting at her collarbone, reading her mad Spanish novels, studying the forbidding ranks of irregular Spanish verbs. In the cabin, in the top bunk, John may be dreaming of his present girlfriend: Shana, Queen of the Greater Yellowstone Ecosystem. The thought of her—long and shy and crowned with auburn locks—claws at my heart.

Thinking of John, I reconsider: in his dreams, he's probably exhausted love and is hungry again. A Rabelaisian dinner, carpets of prime rib and sofas of mashed potatoes, where he reels to and fro, biting with a will. John may not be the hungriest person I've ever camped with, but he's a remarkable eater.

Marty had both appetite and capacity; I'd pack enough food for three and he would eat for two of us. If they both came, a party of three, I'd pack for five. While Marty stows food under his mustache with a wolfish dispatch, John has a soft, almost saintly glow as he contemplates a pot of freeze-dried stroganoff. Quite simply, with a pure heart and a tight belly, he worships the miracle of food and drink, the literal body and blood.

This course of thought makes me hungry. Sleepy, horny, hungry. What a Billy-goat. I should try to meditate or do something noble. Sorry, Thoreau. I roll over and feel the sun nipping my winter-pale behind. The snow is impassable slush, but I'm on standby. This is unpredictable country: it might suddenly freeze. If it does, I'm ready. Still two hours until quitting time. Eat your heart out, midlevel computer bureaucrat.

John wakes up and smells the burritos. He slides from the upper bunk and gazes into the cast-iron pan with delight. He looks around for the boom-box and then remembers that we left it in town, excess weight. His face falls.

"You'll just have to imagine the music," I say.

John laughs, "I put in for another job."

"Where's the job?"

"Puerto Rico," he says. "Tropical forestry. Cheap rum. Swimming in the ocean. The bad point would be the lack of snow. I'd have to come back to ski."

"You could water-ski."

He ignores me. "There's a biologist job in Washington, too, old-growth stuff. Spotted owl habitat."

"Into the combat zone. Do you get a bulletproof green vest?"

"The place is nice. It'd be good, but I'm more psyched for the tropics."

"It might be too warm and easy. Winter is a great rein on decadence."

We eat the burritos with great gouts of salsa. Lacking the hammer of loud music, we start a conversation. When John is desperate, he's a good talker. We talk about whatever enters our minds and go through a certain amount of rum. He describes the weird vegetation on the slopes of Mount Kenya and the black wilderness rangers with carabiners stuck through their ears. I describe the harvest dances at Jemez Pueblo and the molten splendor of their red chili. "To those people, chili isn't a spice; it's a vegetable."

We share a fascination with detail. We take turns describing the texture of places, the plants, the names of rivers, the night sounds, the feel of the air. John tells me about his brother, a guy with a good job who's bored out of his skull. "He sends me tapes of weird, cranked bands. He likes really twisted stuff. It's his release. He must spend thousands on tapes and CDs every year." We talk about the horrors of suburban life, the morass of possessions and debts. "I don't think it's worth the confinement," he says.

I don't miss the background music. It's good to talk. John's interested in sociobiology: how human and animal populations interact, so he discourses on that. We cram the stove and mix rum drinks. It gets dark. The cabin gets too hot. Our talk falters. We run smack into entropy.

This is Punk Physics, a violent simplification. The universal change from order to disorder is known as entropy. Some of this has to do with radioactive decay and the increasing heat generated by ongoing reactions, like solar fusion. There seems to be no corresponding reversal of such reactions. They run their course, the universe heats up and order deteriorates. This quality is also present in our grasp of time, which feels to us both inexorable and irreversible.

The sweet trick of life seems to be to forestall entropy. Living organisms develop boundaries—the cell wall is the most important—and run a sort of entropy exchange across them. In pencil sketch, the organism takes in nutrients that are low in entropy and through chemical reaction loads them with some of its internal entropic burden. Now high in entropy, they become toxic to the organism and must be jettisoned, along with excess heat. But back to the air.

There is evidence that the earth's surface temperature dropped abruptly as the greenhouse effect of a carbon dioxide–methane atmosphere was lost: the grand-daddy ice age took place about 2.3 billion years ago. This ice age may have marked the shift from an oxygen-poor atmosphere to one rich in nitrogen and oxygen. Since then, there have been periodic swings in the earth's average temperature, but they cover a small range: a few degrees. The temperature swings are responsible for the pulses of glaciation which, we are told, are the earth's normal climate. We are living in a brief interval of warmth that is as rare as a clear, balmy day in the Aleutians. Some scientists hold that our biologic sins will trigger greater warming, a problem considering the ever-increasing output of the sun. Others say that despite a temporary warming, a natural shift to glaciation will ensue.

The cabin is murderously warm. John, lizard like, sprawls on the top bunk and goes to sleep. I shuck down to a T-shirt and try to read. Our winter habits have become bad habits. Sweating, I shut the draft on the stove and open the windows.

For we living ones this heat regulation is critical. Our atmosphere stabilizes the solar heat, reflecting it from clouds and haze on earth's sunlit side while simultaneously trapping it in on the nightside, limiting the range of temperatures in one place. Without it, days would be too hot and nights too cold for life, as on the moon. Each of us has a boundary layer of air next to our skins that mimics this effect on a small scale. If you whirl your arm in a sauna, you can dissipate this boundary layer and get a serious burn. On cold days, the dissipation of this insulating boundary layer can lead to frostbite or hypothermia. Animals deepen this layer with fur, while we appropriate their skins or otherwise hold air stationary around our bodies, in wool, feathers, or synthetic fibers.

We also move around. John likes the top bunk, which is warmer. I always grab the bottom one. As the air in the cabin cools, John crawls, half-asleep, into his bag. I shut a window, stoke the stove discreetly, and retire.

We leave the cabin at a respectable hour: it's light but not bright. The snow is set up hard. We skate the flats and cross the bridge. I take this bridge, a rustic wooden structure, for granted. But without it, this would be a very different trip.

The creek rushes under our feet, over ice, between snowbanks. The water is two feet deep above the anchor ice, green and cold and fast. How would we cross it? We could wade across it, but walking barefoot across ice in a fast current would be dicey. We'd have to scout upstream until we found solid ice above the water—unlikely in May—or a fallen log. There's one I recall about a mile up. Then, we'd have to scramble across the icy trunk, skis lashed onto our packs. Nice for the film version, but not for now.

We cross the bridge in four seconds, then forget it. Our wax is working nicely, and the snow is pleasantly firm. Following the bend of the creek to the east, I can feel the cold air draining down from Meeks Lake and then the deeper current of chill from Big Sandy Lake and the Divide, not quite a breeze; more like a slow, deep breath.

Air moves with differences in temperature. It is a truism that hot air rises and cooler air rushes in to take its place. Warmer air is less dense, its molecules bumping each other for more space. So it's lighter. On a global scale, this circulation helps to even out temperature just as a ceiling fan does indoors. The grander movement of the atmosphere is caused by big differences in temperature between high and low latitudes and by the rotation of the earth. Since the air isn't attached to the earth, it lags a bit. On our scale, these lags are major wind currents. All this moving air helps to smooth out the effect of the sun and the earth's spin and wobble, which by themselves would make the tropics and the summers too hot while the high latitudes and winters would be too cold: killing extremes.

As the air responds to temperature difference and spin, so do the oceans. It is the grand sum of these patterns that produces weather. Air circulates both vertically and horizontally, with the vertical pattern owing mainly to simple differences in heat, which cause local differences in pressure.

If you want to see this in miniature, slowly blow a mouthful of smoke over the lighted candles of a birthday cake. The candles create tiny highs, with lows in between. There are corresponding, slower patterns of circulation in the oceans that have much to do with both weather and with the conditions for life in the seas. Oceans have their deserts, their jungles, and their cold seafloor tundras according to their depths and currents.

Surface winds result from the pressure difference between adjacent highs and lows. In the Northern Hemisphere, winds circulate clockwise around highs and counterclockwise around lows, a pattern seen in dust-devils and the gusts around thunderstorms. At a regional scale, this cyclonic movement blends with the earth's rotation to produce prevailing winds.

Now, in Wyoming, in May, our weather is changing. The flow from the northwest that brings the severe cold has weakened, and the prevailing southwest winds of summer are taking over, melting the snow. Soon, we'll have our first spring rain and the snowpack will melt as the streams fill to flood stage.

There are high clouds, thin stringers of vapor riding the high-level winds, southwest. As we climb the glaciated slopes, they disappear, leaving the sky open and bright. We move fast. The snow holds up, our klister wax works and we cross the ice of Big Sandy Lake, seven miles in, at mid-morning. The air over the lake is cold, since five steep drainages intersect, spilling cold air down to be trapped in the deep basin. Even with the colder air, we get hot crossing the ice, since the surrounding slopes are like a great, concave mirror, concentrating the sun's heat.

Airflow over the land encounters the surface friction of forests and the reefs of mountain ranges, around which it forms waves and eddies. This slows the air, just as a stream is slowest at its banks and bed, creating differences in water pressure that swirl the water. In the air, there are swirls—updrafts and downdrafts—along the surface while winds ride faster above. The earth has stratified wind patterns, slower and more complex at the surface, faster and simpler on high.

The wind leaves tracks. Choosing a campsite, I look at the plants and trees for clues: Are there a lot of ground-hugging and cushion-forming plants? Do the shrubs huddle behind boulders? Do the trees seem to lean in a common direction? Is the bark scoured off the pines on their northwest sides? That means strong winds and drifting snow

We climb up the dull blade of a ridge, sweating, trying to beat the thaw. The downslope flow of cold air has ceased, and all is still. Before long, the heated air of the valleys, warmed where the ground is free of snow and dark, will begin to rise up the drainages. Combining with the heat of the sun, the upslope flow will help thaw the surface.

But we're fortunate. We make it to the collector on firm snow. Dropping our packs, we go through our drill, pull the sample, change the bags, split the bag of snow in two, pack it, take the snow cores, weigh them, slap them into a bottle, label it, and stow it away. It goes more quickly when our fingers aren't about to freeze. The sun is strong, high noon. We sit on our packs to eat, unable to resist basking in the warmth. John dozes. I listen to the creek, booming under the ice, thinning it out from below, growing stronger with the day's melt. From the cliff to our north, icicles fall and shatter. From rocks and pines, water drips.

I feel the first breath of the upslope wind, warm and soft. I close my eyes and see the red mosaic of my inner eyelids, like bubbles in cherry jello, a rich design. Everything seems to circle around us as we rest: the swing of the planet, the roll of the seasons, the rising and falling air, the water cycle—snowflake to ice to snowmelt to rushing creek to lake to river to sea.

These powers and patterns are no less vital than the circulation of blood. From a high perspective, imagining the great whorls of air and water, some of our proudest edifices seem to make little permanent sense. Concrete and asphalt keep the soil from breathing out, keep the rain from soaking in. Glen Canyon Dam is a clot in a continental artery, filling its backwater with nutrients needed downstream. The City of Angels burns like a Lucky Strike between the chapped lips of the continent.

So it is said. The sun will become hotter, so the ability of the atmosphere to cool the earth's surface will be vital to the continuation of life. At present, much cooling comes from cycles that depend on the global movement of air, like the water cycle.

The presence of free water gives us the protection of clouds and moist haze, without which the tropics would broil. Some of this airborne moisture nucleates around salt or sulfur rising from the sea and its blooms of phytoplankton, while some of it is the result of the transpiration of plants: largely the tropical forests that girdle the earth's hot zones with green. Clouds increase the earth's albedo, or reflectivity, and bounce a lot of solar heat back into space each day. They also hold warmth instead of allowing back radiation at night.

If we deforest the tropics and also manage to dump enough toxic waste into the ocean to affect the growth of phytoplankton, we could succeed in temporarily harming the earth's cooling plant. The effect would not be permanent in Olympian terms—but the wild climatic swings might last long enough to eliminate not only us but large vertebrates in general. So it is said.

I blink at the other large vertebrate and say his name. "John. Juanito. We'd better go." He sits up. "I wasn't really asleep," he says.

"I wasn't either," I say.

"Nice day," we say at the same time, then laugh. The snow is beginning to loosen, but is still skiable. We make a fast slither down to Big Sandy Lake, mush across the ice, then climb out to the west and start down our track. I can feel the sample, a cold block, low in my pack. I hope it doesn't melt and start to leak. That would mean not only a loss for science, but wet pants for me. We have to guard the samples, keep them uncontaminated and get them out, to keep the chemists happy back there in Oz. I try to recall my reading on the air's chemistry, wanting to tax John with some of it, but he takes one look at my pensive expression and disappears over the hill.

So I keep the chemistry to myself and think. A lot happens in the air: it's a chemical boogie. Incoming solar energy kindles all possible chemical reactions and the diffusion of gases ensures that they take place all over, so the air has the same general composition throughout each layer.

One factor that supports the rapid chemical interchange and diffusion of the atmosphere is its smallness: the air has only a tiny mass compared to the mass of the planet. Another crucial characteristic of the atmosphere is that it supports the action of chemicals essential to life, such as enzymes. Much of our present atmosphere—nitrogen, oxygen, carbon has been part of living cells before. Through many means—transpiration, respiration, decomposition—it is re-emitted in a gaseous state. But it will be part of living cells again and again.

So all the mystical Hindu-Buddhist-Hopi stuff about breath and circles and cycles, all that soft-headed mushy unscientific Redskin mumbo-jumbo . . . it's true. It's a bit short on specifics, but it's a good, sound summary. I take a couple of big breaths. I can smell the sap rising in the pines. I should be reverent, but I'm not. My appreciation for the world is active. I don't enjoy watching it from a state of transcendental calm. I want to get up and dance. Aer, aire, air, aria.

The snow goes to slush. We slop down our track, graceless, falling through into pits of white goop. I catch up with John, but spare him further enlightening comments on the atmosphere. We stand in the sun, eat some peppermints, and are content to breathe. How much do you need to know to do that? The top four inches of snow goes absolutely liquid, so we surf down the last part of the trail. Our skis make sucking noises with each change of lead, splashes with each awkward stride.

The bridge has melted out and the tread is all sand, so we take off our skis to walk across. When I bend down, both legs cramp, so I stagger and curse until the pain goes away, all this much to John's amusement. Then I shoulder my skis and walk across. The creek leaps and races. The ice has broken up, and I see the gold sand of the bottom, glittering and drifting under fast green water. On the other side, I stoop to latch my bindings and the cramps return: more ugliness. John laughs out loud.

We take off our skis again to reach the cabin. We bury the samples—which haven't leaked—in the shaded snow and lounge outside in the sun with bottles of dark ale. We share a bag of roasted peanuts, grinding them with our teeth, washing them down with luxurious gulps. John spills a few foamy drops and rubs them into his bare chest.

We'll ski and slog out tomorrow, starting early to get out on the frost. After the long cold season, these first warm days make us coyote-lazy. "Wish we had the boom-box," John says about every six minutes. "No life without music." Waving my ale bottle, I make up a little song:

> There once were two rangers, Demons for dangers,
> Who risked all for Science and Sin.
> They brought home bananas, to Linda and Shana,
> Who welcomed them both home . . . and in.

John salutes and drains his bottle of ale. We dig up two more, shining wet out of the snowbank, and uncap them with my Swiss Army knife. To be barechested on this warm patch of ground, under this particularly bright sky, to be guarded by a thousand square miles of impassable, gleaming slush, is to know rare bliss. I can smell the lodgepoles waking up, their pitch pushing out to the green point of every needle. Like newly-awakened bears, we loll, aromatic and hairy, in the sun. John hums scattered lines from a song in a drowsy voice. Every time I start to talk about the atmosphere, he replies with a global yawn.

The Atomic Nature of Ma

Lori Anderson

figure 9.1: girl model with model molecule*

she is made of stardust
the carbon, the oxygen
the other atoms that make her
body originated in the deep
interior of ancient stars
long since exploded
you are she in me

for each breath you offer
every person worldwide takes
one of your exhaled atoms
we are breathing each other
many atoms in my lungs
were once in your lungs
the lungs of every person
who ever lived we are
the same atoms we borrow
by breath by sweat

your grandmother's last breath
is Adam in my belly
take him from me this Eve
mount me, mouth me
until I have your granddaughter's
first exhale left in my right lung

figure 15.13: sleeping pigs side by side

pigs have no sweat
cannot cool by vapor
they wallow in mud
hog breath rises rises

*Note: The figures referred to in the poem are from Paul G. Hewitt *Conceptual Physics,* 5th ed. (Boston: Little, Brown 1985). The New Testament reference was paraphrased from a New American Standard Bible.

expands chills coalesces
chance collisions we
fly through their cloud
we rut they rut like dogs
each intake we be swine

your grandmother's consort
is a boar in my blood
make her from me love
mount me, mouth me

Mark 5: 1–13 man from the tombs with an unclean spirit

he had often been bound
no one was strong enough now
his shackles broken in pieces
by night he gnashed himself
with stones implored God
the Adam-in-him already
within our Jesus asked
his name: my name
is Legion for we are many

with one breath of the Lord
unclean spirits entered swine
two thousand rushed a cliff
drowned in that very sea
the sea that clouds our exhale
for each breath we hold
that very man from the tombs
who breathes one of our souls

life lunges toward the unclean
Legions' atoms in our lungs
like Jonah in a whale's belly
ride the tide of conversion

figure 10.6: the stretch of the spring

an object subject
to deforming forces
undergoes change
depending on bonds
the order of atoms

if molecular vibes increase
molecules will shake apart
will wander taking shape
ordered by their container

each atom rises rises
expands chills coalesces
chance collision we
were all hog breath
in the six hundredth day
of Noah's life we were
all fountain all ark
vapors of the great deep
burst open we are every
beast and every creeping
thing we are two of all
flesh which was each breath

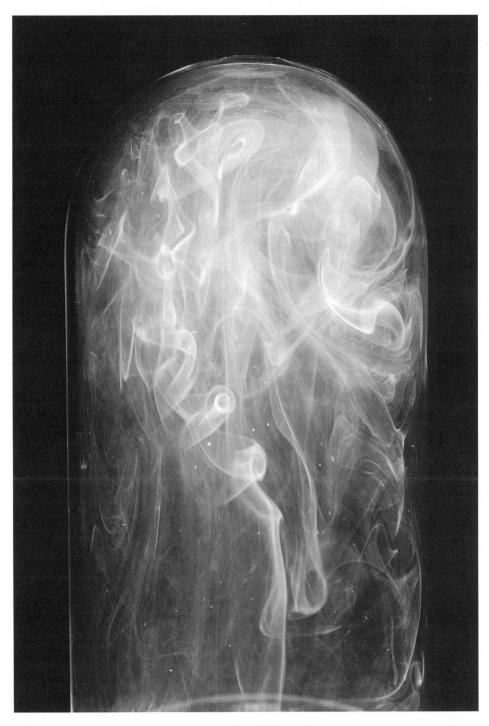

Susan Derges, *Aëris,* 2001. Courtesy of Michael Hue-Williams Fine Art, London, and the artist

Surveillance

Edie Meidav

Into the airless night, he is working, and when he forgets what the mission is about, he raises his head and looks at the photo he had pasted above his desk, something photographed just two years ago but even then made to look old, already fading. When first posted, the image had possessed sufficient gravity to keep him motivated. He had been right about this: the gravity had adhered even when motivation has faltered. The image cannot be erased. It is a double-headed, fire-breathing monster, born of the moment when everything first started falling apart at a rate beyond what most people could comprehend.

Everything about the image satisfies: it is as responsive to his mood as an attentive lover. In its unblinking gaze upon him, the photo knows him, over these past two years, far better than anyone else.

He goes back to work, without letting fatigue drag at his labors. When tired, he has become an expert at narrowing down into sheer unbreathing focus, sharpening his listening, hearing patterns. No matter how tired he is, his fingers stay light on the switchboard, ten bird beaks pecking impatiently into the next conversation, the one beyond that, and the unimaginable one that lies far beyond all that.

His greatest skill remains this: unlike his fellows, he had from the beginning been able to avoid too great an attachment to any one conversation. He can listen to the conversation waiting to happen beyond the horizon; this trait, among others, has singled him out as someone of great worth. Detachment coupled with focus: he can scan all words, no matter their languages, within the ether of a half-listening; he pays attention only to the hints dwelling at the edges of conversation.

True, all the fellows possess more than average faculties—retention, memorization, speed, acuity, linguistical analysis. So much had been vetted before the training. Much like the enemy, superior men, they live anonymously. Some have families, yes, and wives, and yet if people ask, as they would, these attuned men were to say they work in a boring civil service capacity, paper pushers for the government.

As much as they had stepped forward to serve, they had been met by scrutiny, to the extent that a swatch of arm hair from each recruit had been posted to an overseas lab, location unknown, for a highly rigorous chemical analysis.

Unusual exposure to mercury, lead, cadmium, and benzene, his particular swatch had said, but the Squad had needed to lower the bar. It was likely that someone upstairs had sensed his particular promise; they had let him in. This had all been explained. There had been no sloppiness. It is, in part, the cleanliness of the Squad, the great gray building, that shocks newcomers with its clear statement—though Operations had once been caught unaware, every arsenal of modernity would now be pointed at the threat. Superior aim would lead to a superior overcoming. So it is that every line and angle of the Squad screams victory; even the two phalanges of walkways toward the building form a tight **V**. Yet you cannot find the idea of victory in any training manual, mainly because Operations abhors all paper trails. Hence, memorization: this is what counts.

Two years later, he still finds pleasure in his job. It is akin to what someone who collects safety pins in circles, or paper clips in a bundle, might feel. He sweeps together the stray ends of discourse, pauses, air crackle, ordinary tags of dialogue, into the kind of heap that might allow a greater code to emerge. *There is always one conversation behind all other conversations.* This is the premise, the origin and destination, the first thing they had all learned.

Another part of his acumen comes from this: even from the beginning, he did not find much to speak about with his fellow surveyors. Eighteen months together in training, learning about listening and psy-ops: he had been neither rude nor mute. Just self-contained. The rest of them, overawed by the cinderblock eminence of the Squad, which would one day promote them up its floors, needed to fill moments of relative leisure with an outpouring of chumminess. They needed to replace all the words they had heard with sounds from their own mouths.

At breaks, the other recruits would compare notes; in the refectory, over meals, they chatted of inconsequentialities. They lingered over previous travels, misconceived ideas and lives. All the while, he chose to sit apart, stroking his smooth cheek, staring at the white walls; he ate his meals methodically, mixing peas and carrots in appropriate quantities, varying amounts with each mouthful. If he had time, he would pace the reflective halls, stare blankly at the official bulletin boards, come to locked doors, listen to his own footsteps and those footsteps beyond. He would turn back, pace more. He knew the Squad would work its magic. If not at first, surely after the first year, they would all became true listeners. Soon all would display a keenly honed professional silence. Given this destiny, why should he be interested in speaking with any of them?

He did not stay apart in order to outperform the others in surveillance: yet it was true that silence raised his professional IQ. Sustaining the truest silence, he could narrow into the earphone's conversations. And so, though not especially gifted at any one of the three core languages, he surpassed the others with his focus. It was this focus that made him the object, if not of admiration, at least of envy. In the elevators with others, he was granted a kind of bubble. No one asked him anything; he returned the favor. They knew his name was Christopher: that sufficed.

Before the catastrophe, before he'd signed on to the cause, Christopher had lived something of an unruly life. But the catastrophe had changed all that. Like many of his peer recruits, in a general sense, he had battened down the hatches—signing on for the cause had been only the most specific gesture.

Fitting with their mission, after the training, as Christopher had predicted, the recruits found all their chumminess disbanded. The magic had worked. In the second year, they were isolated, as if isolation were an honor and privilege: each was given his own area of a floor, to which he possessed his own key. The only oversight became then conscience, and the reports submitted to the female supervisor. Christopher learned to appreciate the system's honor, one that he had married. Without the code and mind-numbing drudgery of the endless hours of surveillance, where would they be?

Walking the crowded, unthinking air of cities: open walking targets for the threats and spores of madmen. Far better to be enclosed, purely listening, breathing the vetted air of others. In midnight hours, he grew especially alert to the danger of abstraction, when his whole being would start to float into a roving ear, high above the earth and its conversations, all nuance and beastlike suspicion. He thought, at those times, how truly thin were the veils of civilization.

If he gets too tired, there is always the danger of this floating.

The danger being, at such times, that all conversations become equally significant or equally innocent. At such moments, honor demands that he call in for a break. A certain humiliation rests in this, dialing his chief on the Squad's tenth floor, receiving the confirmation, retiring to lie like a fallen wasp on a cot next to his cubicle. Listening to the delicacy of the white noise, strapping on the musty green eyeshade, or just watching the clock, its blue digits piercing the air with a hostility far greater than his workroom's clock.

He tries not to take these fifteen-minute breaks as often as his peers, and in his first few months was so successful at this that he was allotted not just his quarters but his own floor, one with a solo latrine where he never has to cross anyone else's path. This privacy has become a new sort of lover, as tender and cosseting as a mother.

Into this system, set up with the diligence of inertia, a new recruit is to enter. The fact of this new recruit would not have made anyone happy. The chief's voice—in which one could practically hear the frowning brows, the eyes goggling over the wire-frame glasses—had announced the new recruit, and yet at the tail end of the announcement came a certain softening into delight, as if his chief must have guessed that he had been considered a little barbaric by Christopher.

"You won't have your turf undisturbed anymore," the chief said over the intercom.

To his chief's gleeful tone, Christopher had nothing to add.

"The training program has been streamlined," the chief explained. "We've had to do it. It's the new policy."

"Why?" Christopher finally asked. He had found it best to simplify.

"There've been complications," the chief had said. "It would be a good thing for everyone concerned if our new recruit can take over"—he actually said that, *take over*—"a portion of your switchboard."

Christopher, listening, cut to the chase. "My figures are falling short of the unit's expectations?"

Into the intercom the chief let out a little sound, some combination snort, click, guffaw. Quickly aborted. "Hardly. This will help the new man learn the ropes. Back when you first signed on, we didn't have this need. It's necessary," his chief finished in a kind of musketry. "We're expecting full cooperation."

After that, Christopher could almost hear the sound: the raising of the bushy eyebrows, the stare ugly over the glasses, the admission of no rebuttal.

Hadn't he signed on for this clarity? The command, the obedience. Even the eyebrows. *This is all for the good,* he tells himself, *the mindless exertion.* When the digital dial reads ten o'clock, the blue digits crystalline like anything else in his cubicle, if overlarge for the small clock in which they live, he welcomes the cue and removes his headphones, all their simpers and hisses.

It is true the new man has not yet arrived. Christopher tells himself he is not about to lose his job—the only job he is good for anymore. He cannot quite remember his previous life, though shards of it occasionally come back. There had been a house, naked branches. Various hands had held his, had let him go.

Without his bidding, a shard does comes back now: the smell of autumn air, burning leaves, the first cold tickling one's wrists, one's nostrils. He shudders it away. He does not yawn, nor does the recycled air of the compound bother him. The steel vents at one corner of his office speak of a connection to others, though they hang like sadly modernized cow teats. They dispense air and suck it back up, extracting the carbon dioxide breathed through the lungs of tense men. Sad cows, yes, perhaps, but they shoot air out an unseen back alley at the rate of seven hundred units an hour. The slipshod construction of the vents, their screws and bolts showing, such a vestigial reminder of community and breath makes him believe the ducts must have been an afterthought in the design of the Squad. Surely more deliberation could have been put into the vents.

But he likes the slight chemical flavor to the air that comes through them—a combination of ink, tar, scorched milk cartons. The scent of the first day of school, one's head on a scoured desk, inhaling the freshness of bleach. He actually thought the compound would be a truer entity if all the surveyors were to breathe less. *We are all just intelligence gatherers,* he thinks; *what does oxygen have to do with intelligence?* In his more fatigued moments, the pumped-in air seems a punchline, provisional, like the laughable nature of the Squad's petty hierarchy of privileges: its solo latrines, its cots and white noise, its calls to the chief, all contained within the memory of the sad cow teats.

For what it's worth, he tells himself, he breathes what they breathe on the tenth floor, where the chief works.

His first months, he'd had little slack in his being. He'd been shorn for the cause, streamlined, had spent no time contemplating oxygen. He'd tightened his belt so he could breathe less and focus more on some of the world's conversations. Five hundred lines of conversation. Quite recently, his entire being had coincided with the greater mission.

In the last weeks, however, he finds it harder to stave off fatigue. He calls in for back-to-back breaks and waits for dreams to overtake him. He bolts awake and stares at the persistent sky-blue digits of the alarm clock.

Something unsolvable has started. He can feel it: parts of his brain are not exactly eroding but shrinking, as dead as withered potatoes. Meanwhile, the part of his mind devoted to discerning codes in speech has become so alive it crackles, blue-electric, pulsing with an unquenchable thirst.

It is not that he wishes to catch perpetrators, past or future: not that he wishes to be rewarded in some blaze of glory. No. Rather, the longer he stays in his cubicle, eavesdropping on his allotted five hundred telephones, switching from one conversation to another, the more he needs evil to arrive. Evil will justify the means. It is not long off. He can feel it.

Unfortunately, the night he is to train the new recruit, he stumbles on an unusual conversation.

Between two women, he thinks, unless one of the conversants had affixed a gender scrambler. On a phone line he has known for more than a year now, one used by a curt impatient man named something like Ravi, though this Ravi aspirates the last syllable; the name could equally likely have been Radi, or even Robbie, the emphasis being on the Ra.

Two women, however, he is fairly sure; one of them uses Radi's line; and their whispers sound emphatic, happily in one of Christopher's better languages.

At first, he believes he has found a jealous quarrel. As he increases the volume on his phone, tipping into the conversation, he lowers the bass so that their voices, one upper-upper-treble and one merely alto, find a perfect fidelity.

Were they speaking harsh words of endearments at great risk of being caught? Guilty, he looks around. No one is in the room with him. Unwilled, his gaze catches on the photo of the catastrophe, its left corner already starting to curl off the wall.

"I can't tell," the upper-treble woman is saying.

"Please," says the alto. "You're being ridiculous."

"You could try to understand my situation."

A great silence follows, enough for an eavesdropper, guilt clouded, to think the connection had been severed. Impatient, he presses the wrong button, only to find himself within the conversation of two men whose greatest concern is whether to invest in hog futures. Hog futures! Christopher undoes this connection, plugs in another, plugs back to the two women.

"I'm not saying your feelings are unimportant," the alto is saying. "But Radi—"

"Please. Not his name on the phone," says the upper-treble woman.

"You're being silly," says Alto.

"Protecting my husband?"

And then Radi's wife must realize the gaffe. She sniffs at herself. "Call me when you have something kind to say. I need—" and then slowly at first, up and down, and finally with a great slam, her connection is cut.

Perhaps the phone had been slapped out of her hand. Perhaps a man had entered her room, insisting she cover herself. Perhaps even as Christopher sits there, the woman with her high but slightly scratchy voice, one like pearls rolling within a silk sac, swathes herself into a giant bandage. She breathes heavily. She has created for herself a heated zone of personal inquiry. Does she question living within a state of service or refusal? Only the man's eyes have been allowed entrance, only his hands have been granted knowledge.

Christopher tries to return to the two businessmen talking of prospects for an insurance company. They talk of annuities, slab foundation. An uninsured building. There are windhole entities to be seamed, cutlass surprises to be bent, an actuarial table that has been fixed. He takes off his headphones.

Just then, of course, his employer enters with the new recruit about whom he has spoken with such demonic satisfaction.

"We scoured the lower provinces to find a graduate like him," his chief had said the day before. "Our new man is schooled not just in our seven languages but also in the lesser triad and in the far-off couple. This makes him one of our most versed intelligence experts. We trust you will hand-hold. Just the first two weeks of his stay. He'll have quarters on the same floor as yours. For now, we are moving the Irishman upstairs to be closer to management."

"What is this?" Christopher had stammered.

"This is our attempt to streamline the training process. Not every day do we get to find such a promising recruit."

"You mean this two-week whatever is going to supplant our own two-year training?" Christopher had asked. He had not meant to insubordinate but was unable to help himself.

"Strange times call for strange measures," his boss had said. "It is the wicked who make history."

"No," said Christopher, quoting someone, "it is historic times that turn us all a little wicked."

"It is anyway how we are doing things," replied his boss.

Left alone with the new recruit, Christopher swivels toward him. Yet another with eyes only recently weaned from the milky blue of a computer screen. These people have the depth of a vacuum but become necessary in strange-measure times.

"You know the Psy-One system?" asks Christopher, sighing, ready to launch into his explanation. "This is a project in which the repetition of certain words is tracked by a massive centralized computer—the statistical frequency of repetition begins to predict the timing and locale of the next attack."

"I just got trained on Psy-Two," says the new recruit, unable to wash an equally milky satisfaction off his face. "The newest one? My name's Christopher as well. Can I call you Chris?"

"I'd like to keep my own name. If you don't mind. I began the job with it. It's my name, after all."

Luckily, the new recruit doesn't disagree, and this constitutes enough of an agreement for *him* to become Chris.

Two weeks, two weeks in which they will have to work alongside each other. It is bad enough when young puling Chris, as promised, turns out to be a hurtler, as they call it—one expert at tracking conversations, at feeding bytes of information into Psy-Two. The cad ends up being the one to train Christopher. But that he must do this with a slight sneer hard for him to surrender is insufferable.

Young and needing only an infusion of bubbly caffeine, the new trainee has already found one suspected attacker. He has managed to construe an electronic trail that probably stops only a few degrees away from threat number zero.

Christopher feels his lair invaded. He wonders, too, not idly, if he is about to be replaced. What if he were abruptly to be sent into the world of randomness again, where the wind shakes leaves on a tree without any real discernible order, where a woman passes you on the street and her scent brings up all sorts of unbidden memories? He hasn't been slacking.

But he does begin to take chances.

He punches his keypad to light up when lines connect the two women—Radi's wife and Alto. His impatience when these light up—and he cannot listen—is such that he could have swiveled Chris's head on its axis to have gotten him out of the room. He would like to dwell in the labyrinthine center of the women's talk, breathing in its swirling fumes, its histories, its urgency.

Yet only in the beginning of the second week does he have the good fortune to see the lines brighten while Chris is on a refectory break.

How long will Christopher get to listen this time? Perhaps the new recruit merely sits in the cafeteria, sitting in a corner by himself and staring into his own private horizon. No doubt young Chris would be blowing massive bubbles from the free chewing gum that is everywhere to be found on the compound, massive containers of it, since bubble gum is the great perquisite, what their training had called an effective anger absorber and anxiety reducer. Christopher understands this: the air is first encapsulated, and then popped; air remains fully under one's control. He himself has recently subscribed to this idea and now is never without a cache of oddly pink-flesh-colored gum wrapped in the operation's insignia, a perk ready for him.

But with the two women once again connected on his line, he stops chewing. The second conversation of theirs that Christopher has managed to enter, and it does not bode well.

"I take pains to not let anyone know about us."

This said by Radi's wife.

"But you do not follow necessary measures," says Alto. "For all we want to do, you have to give up more of your defense."

"You want too much of me." Radi's wife sounds edgy. "You want me to give up everything."

"Carmella!"

"Shasia!"

Or these sound like their names. Christopher pops two pieces of gum into his mouth and writes down Shasia and Carmella with a sharp-nibbed pen onto his daily ledger. They have hung up already, but he lingers on the idea of Carmella. He is drawn to her treble nervousness, this wife of Radi. The other one, Shasia, is pushing forward some goal. He envies them: their days spin forward, they have purpose and connection, yet their goals remain a mystery, waiting for him to understand

"I will need you to issue performance fulfillments on a more regular basis," the chief says to him, meeting that afternoon in his office.

The chief himself is quite expert at blowing bubbles, giant irreproachable bubbles, accurate in their sizing of air and pop objectives, and as he speaks, he stores the gum in a lower and especially pouchy part of his cheek, so that Christopher imagines he is chewing on some superior spy instrument that has helped the boss achieve his presence on the tenth floor. "You have five hundred lines you are monitoring. For obvious reasons, I can't keep an eye on how well you are monitoring them. If I were to do that for everyone," and he waves his hand, indicating the Squad could easily crumble into a half-completed circle.

"We are in a time of crisis. Our usual methodologies have been, one could say, confiscated," his chief finishes.

Christopher twiddles with his ledger. Inside are the two names, pulsing. Shasia and Carmella, Carmella and Shasia.

This meeting has been, after all (hadn't he known it?), mostly inevitable. Some part of him wants to open up the ledger, as if he were a flasher. He wants to reveal to his boss something beyond the pretense of their mission. He wants to reveal the greater unclassified humanity taking place in his ledger, far beyond what the external squadron might know about meetings near public restaurants, private bars, bugged park benches. Beyond the ripple effects of his reportage on conversations. In his work, Christopher has never led anyone any closer to the source of all danger.

Or perhaps he has; perhaps Operations has not kept him informed. Yes, he has indexed various speakers as being threats A through Z. Yet nothing, so far as he knows, has ever come from his efforts.

There really does not seem to be much point to it all.

When he had signed on, he'd felt the importance of the mission, and had gladly sacrificed two years of life to the greater good, the societal hum, the workings of freedom. Now, with the two years stretching backward, he feels he has patched together a giant balloon that may or may not ever arrive at its destination. Of course, he will never see the balloon; he can never attempt to pilot it; the know-how of his nation may not convert his toil to the good. Good, however, is the only misty ether left to believe in.

"You have to trust the importance of the mission," his boss says, before dismissing him. Poppingly.

"I trust in the importance of animals," says Christopher, thinking of cows, realizing from his boss's quizzical look that what he has just said makes no sense. He has stopped listening carefully to his own words. All that is part of the dead potato.

For the gatherers of intelligence, the only collective common denominator, functioning for all of them, is their suspicion. *One day,* Christopher thinks, taking the elevator down, *I will become suspicion itself. And nothing more than that, nothing less.*

That afternoon, before it fades unwitnessed into evening through one of the corridor slits, the new recruit is to be taken to the Irishman's level upstairs. The Irishman is to be booted, for reasons too terrible to breathe.

This is what the recruit has indicated. "Upstairs they'll call me Christopher," he says, a final taunt. "If you don't mind?"

But Christopher, the real and first Christopher, is glad to be restored to his solitude. He puts on his earphones while still saying good-bye. The recruit leaves him, finally, alone again. And his special lights have gone on. Amid the bustle of all the urgent conversations taking place, in the panoply of languages, Christopher is successful in zoning in on Carmella and Shasia, today eloquent in their silence.

"You will come to me one day," Shasia finally says. "It is not exactly revolution I am demanding."

"It goes beyond revolution. This is what you fail to understand."

And now he is sure of it: the phone must have been struck from Carmella's hand. There comes a distant cry. He hears what is perhaps a chair scuffing on the floor above him, or it may have been elevator pulleys in the wall. He wants to understand.

"I am sorry," he whispers.

Shasia is the kind of woman you could imagine feeling after she had passed by you in a crowd, her sleeve grazing yours, her presence one that probably never goes unnoted. He realizes he'd forgotten to turn on his mouthpiece. There is a slight click as it buzzes on, the mouthpiece he has used only when communicating with his chief.

He stays hunched over the phone. "I hear you," he whispers again, this time with more articulation.

Nothing returns to him in the dreadful transcontinental void. He notes a dusty wind. Could he be imagining it? A wind that has begun to curl from the air ducts in his room, clouding over the photograph of the catastrophe. The pipes within the Squad walls have started to knock, to rattle: one thing he is sure of is that the photograph will not fall: he had affixed it originally with standard-issue putty, called flesh-colored by its Malaysian manufacturer.

No one says anything for at least a minute. For a second, he wonders whether he did in fact speak. Then he becomes sure of it; they are both waiting for him to speak again. The two women are waiting for him. But what can he tell them?

"I am your enemy," he says.

They do not respond to this.

"I am something you have been warned about."

No one adds to this.

"But I would like to help."

More of the same quiet. He notes the air from the ducts is as rancid as a godless desert, round perfect brown clouds gathering in on themselves. It is as if each word he utters brings forth another cloud. Has he upset the ecology of listening?

"You want connection—it is a worthy thing. I wish I could escape, come help," he says, coughing.

"No one can give us anything," says Shasia finally. "We are beyond hope."

"I would like to help you."

He is a penitent, knees on the floor, chin on folded hands on the edge of the desk.

"You are beyond help yourself," she says. "We are hanging up on you."

"Please don't," he coughs.

"We must."

"What is must?"

"You can have no idea of us."

"I can," he says, going hoarse.

He moves to sit on the floor, dragging the switchboard to the desktop's very edge, where it stands the risk of toppling. "I want to understand," he says. "Won't that help?"

He feels cozy, connected. A tunnel is being dug toward his heart, a redemption, as if he has been waiting for this moment for a long time.

"Carmella, we are hanging up on him," says Shasia.

"No," he says. "Please. I would like to help."

"Good-bye," Carmella whispers, and he imagines she is wistful about this.

"Good-bye," he whispers back, just as, the last thing he will see, on the screen behind him, floats the terrible message. It is not that he should gather his belongings and report to headquarters. And it is not that a bomb is being deployed on the two women's homes. But it is something far worse, what he has never been able to imagine until just that moment: They have found him implicated.

IV

"Blow, Winds, and Crack Your Cheeks!"

Jaanika Peerna, *Lines in Silence* #7, 2001

Divide Winds Charge Ultimate Price

Reg Saner

It's snowing sideways. My skis leave a trail blown over in minutes. Good.

"Let me get this straight," says the mind's other side. "You like mountains best in December and January? When there's nothing up here but sudden weather and snow?"

"Yes."

"And another thing. Why so keen on skiing alone? The obvious motive is fear, isn't it? Alone you're afraid, a little. You know you are. Admit it. Isn't that the attraction?"

The mind's other side has that twentieth century quirk of supposing base motives are truest. Hence its sentences beginning, "Whereas what you *really* . . . " No use replying that a taste for the out-of-doors isn't always juvenile. Against one's own brain, what defense comes to more than a shrug?

"Seems to me, " says that brain, "what you're really fond of is bad news."

"You mean the truth? And 'really,' isn't that what we're here for?"

No end to self-skepticism. Out of its duet for one have evolved the power and tediousness of our species, so my in-progress last word to myself on skiing up here alone is only part of the confession: "Colorado's back country in winter is tricky indeed. No secret." But when we look honestly at this world, maybe the greatest risks take place inside us.

Last week, near timberline, I skied into a high country made of exactly its own black islands of towering fir, exactly its own snow-cornices curled fantastically over thick ledges, exactly its own snowfall of fat flakes. Because the grey sky was stalled, as close to motionless as Colorado wilderness ever comes, its snow floated down almost vertically. Rare. No wind, just air leaning slightly; gentler than breath. Against dark fir trees each flake descended apart from the others, a tuft of clumped crystals, a once-and-for-all of intricate stillness made visible.

Any "This, Here, Now" so entirely taken with being exactly itself can't help arresting a lone skier, just as any mind that arrives there takes one look and stops mumbling. Stops cluttering itself with thoughts. Hasn't a name, isn't anyone. Becomes what it hears: mountain snowfall in which silence ripens.

But today, skiing Glacier Gorge, weather tries to tear off my head. Failing that, wind settles for my knit cap—hurling its burgundy wool like a limp blossom into a naked clump of dwarf willow. Fetching headgear from willow twigs while wearing skis seven feet long will improve anyone's sense of being the Unknown Marx Brother. By the time I get my cap back, snow blown into my hair has already clotted, to go along with the crisp feel of January's edge at the rim of the nostrils. Oh well. I'm out for whatever. Today, whatever is wind wind wind.

Owing to chill factors, mountain wind can seem triple-sinewed, hyperactive, fanatic. Isn't it this minute dilapidating the mountains themselves? Even without ice for an ally, it could do the entire job, and has, on summits ground down to prairie before these mountains first lifted. A chip or sliver here, a grain there. Wind and rain have all the time there is. In the central Sahara are mountains that wind has brought down to the humility of floor tiles. Their "mosaic" remnants stretch as if hand fitted—like wind's anagrams, or a code without messages—level to every horizon. Given air's origins, that seems ironic. When stone invented atmosphere, how could it have guessed the strongest force on our planet would one day turn out to be air?

That air with its weathers now intends to level the Rockies? Oh yes. And shall. Nothing can stop it. Meanwhile, I suspect that our part of North America may not contain upthrusts more naked than these granite faces forming headwalls in Glacier Gorge, whose hierophantic jut rises ahead of me: the cliffs of Longs Peak, of Pagoda, The Spearhead. Against them our life spans have no chance at all, just as their own slab-sided ramparts, against wind, have none. This too is where gods live. Minor deities, it may be; forces dumbly, humbly immortal. But godlike in one thing at least: not knowing how to lie.

Near Mills Lake, tiny white mushrooms of snow pique my curiosity. Some coyote or fox must have trotted past, compacting snow just enough. Now wind undermines those tracks till each paw pad stands pedestaled, two inches above surrounding crust. Amusing. I've heard of wind blowing bark off the trees, blowing chickens into the sea—white chickens, and in the Hebrides, as it happened—but never of this. Yet the likeness between these ice mushrooms and pedestaled stones in New Mexico, Utah, and Arizona bespeaks inanimate nature's versatile monotony: two or three ideas, varied perhaps less than a dozen ways.

Altered states is one. Mills Lake is now a long pool of fluid gone rigid. Over its milky, wind-polished ice there's a skinny line of ski-packed snow that, like the paw prints, managed to stick. All the rest, blown clear. I choose the lake's locked edge, its snow so wind packed you could quarry or saw it—a crust my metal edges barely incise.

Wind wind wind. "Restless" doesn't touch it. The sky, however, has cleared except for rags of fast cloud which quickly veil the sun, quickly strip it naked.

The day that wind slowed Ruth Magnussen till it killed her, I was a few miles east, two thousand feet lower in the same weather. My friend Ron and I were trying to ski up an old mining road to Buchanan Pass. We did OK till the twisting, forested road entered a clearing. There wind had a straight shot at us. We leaned forward, fought with it, tottered comically in it, made headway more and more laboriously against it. I remember how angrily the bee-sting of snow pricked my face; and how, shoulder to shoulder, Ron and I heard each other's torn shouts as noise, not words. In fifteen years of our fairly pedestrian mountaineering together, we hadn't let weather deflect us; nor have we since. That day we did.

Next morning under DIVIDE WINDS CHARGE ULTIMATE PRICE, I read how in that big blow, Ruth Magnussen's climbing party had let her lag behind, then on their descent from the summit of Mt. Alice found her "not moving." And I knew it wasn't just any wind. It was *that* wind. Even so, the incident puzzled those of us who knew that her group was not only experienced, but included biologists. They of all people.

We like to believe in the avoidable. We? Not those who describe this world as machinery. The Marquis de Laplace boasted that if he could know for one moment the exact position of each particle in the universe, he could predict everything from there to the edge of infinity. If. A very big word. Needless to say, he lived in the era of Newtonian physics. In any case, it's an omniscience most of us feel blessed in not having.

To see wind purely as a jostle of gas molecules deanimates even its most baroque whims; makes them mere products of solar rays heating Earth, and of our planet's rotation. Which they are. But if our future is physical laws, have we one? We'd prefer weather whose quirkiness is odd as our own minds—random, capricious. Otherwise, strange to say, we can't take ourselves seriously. The world wouldn't be real enough.

If meanwhile the laws of astrophysics have a steely, predictive glint to them, living matter is—happily—a truant to prediction. Our biological past is an unbroken record of the slightly off-center, the slightly imperfect. Praised be imperfection for it! If we owe ourselves to evolution, evolution owes itself to that one continuous, omnipotent glitch. Mere physical laws might have wanted living stuff to repeat itself with cookie-cutter perfection. Nothing doing. Only the flaw, only the glitch in repeatability seems flawless. Its perfection is that nothing repeats, not exactly.

It does so almost. A nudibranch designed to recur without bumps somehow gets one. The bump somehow gets passed on. Imperfectly, though, because that bump gets another—a bump that its encoding hadn't been encoded to ask for, just got. And that bump, another; and so on, till bumps grow ruffles, ruffles grow gills, gills grow brains, and brains blunder their way into eyes. We ourselves are anthologies of birth defects that proved workable—like every creature that is—because life is "error tolerant." Hence "junk DNA," a nucleic acid that neither promotes replication nor deters it, just floats uselessly around in our cells. Through the necessity of chance—and perhaps through oddments like "junk DNA," whose potentialities at—some minuscule difference may begin a current that "flows where it pleases to flow." Thus the great ash tree Yggdrasil ramifies. Thus the fruit eaten by Adam and Eve sows within living cells its escape properties, their enabling imperfections.

A fortunate fall, that. It released us from the tyranny of perfect replication to the creatively flawed: nature alive and ever so slightly askew. Whose creatures we are. But why *these* natural quirks become laws, and not others? "Among worlds that never happened," I often ask myself, as the soul of certain chemicals made mortal, "why exactly these limits and mysteries?"

Air, for example, has one-sixtieth the viscosity of water. But mightn't wind have run as rivers now do, in fixed channels? Streamlines. If you didn't live near a wind's flow, you wouldn't get any. Meanwhile, water in its usual state works that way already; a wind too heavy to fly.

On the other hand, wind isn't wholly amnesiac. Lacking fixed channels it shows preferences verging on habits. Winter to winter, my return past particular snow-drifted ledges, boulders, and overhangs reveals air-carved forms I remember because wind remembers how it made them. Each year, it reshapes them so much the same way that I've come to rely on and use for landmarks their mimed turbulence—where otherwise your skis are the trail.

I know too that although I'm now damp with sweat from kicking side-steps up the deep powder of the last steep ascent over a frozen watercourse, there's a most chilly blast expecting me at the lake's lip. No matter how extravagantly the here-there-and-everywhere winds lurch or wheel, one special air torrent, assigned as if by decree, pours down over that lip like the waterfall's winter ghost. Ordinarily I'd be braced for it, but the ascent has so belabored my breath I forget. As Colorado altitudes go, 10,600 feet is nothing; however, what with skis, ski boots, a winter pack, I notice my lungs have lower standards. Over this wind-crusted snow just below the lip of Black Lake, my metal edges barely keep me on the slope; but the sole risk is looking silly, sliding helplessly down about a hundred feet and having to reclimb the same snow.

Then it attacks. Air's avalanche. Like thunder, a wind mightier and older than life itself suddenly lunges past me, hurling, screaming, charged with snow billows like stampeding herds of white bison. My wool headband isn't enough, needs help from my parka hood—which seems entangled by pack straps. I fumble frantically, because the wind-chill is instantly serious. With that hood finally yanked over my head I watch Niagaras of snow boiling past. Between me and the black-green of fir forest below the lake's glacial shelf I see the frantic writhe of white crystal, as of some creature, spasming. I feel I'm inside a comet, robed with its broken ice and cold gravel. In such wind even granite might wish for life enough to die, to be killed, "Once and for all. To get it *over!*"

No such luck, not for granite. And that may be the sadness of stones. Lives briefly intense as our own can at least expect an oblivion—endless, durable—which the life of granite is far too dim ever to hope for. As the berserk current abates to just weather, I realize it was the wildest turbulence I've ever been out in.

Anyone new to Black Lake might turn back here and now; however, long acquaintance has taught me that by skiing another fifty meters further, apparently into the teeth of it, I'll get clear of the worst. The lake's outlet being a granite gap in high cirque walls, its breach simply funnels cold air falling off the peaks; and, as from a funnel, those currents stream fastest at the spout.

In the great cirque above the lake's own far smaller cirque, it's true, vagrant wind cells lurch around like bullies. Exposed fir grow all twisty, lower to the rock the higher one goes, till—past timberline—their final clumps grovel and surrender completely; wind's abjects.

Roaming our planet, the same wind keeps changing its name to suit place and mood: zephyr, breeze, gale, blue norther, chinook, tramontana, twister, cyclone, tornado, hurricane, williwaw, mistral, sirocco, harmattan, monsoon, typhoon; the trades, the westerlies, the squall lines of easterlies; varied sounds naming air's seasons, its motions, contortions.

Another sort of wind word, *ventifact,* appeals to me especially, not as one of wind's names but as a term given certain of its victims: sandblasted pebbles and small stones on which it imposes polygonal shapes, cleanly sculpted ridges, oddly pyramidal hints. Stone litters thus abraded—whether found on the high barrens of Outer Mongolia, or across central Saharan flats literally paved with miles of the things, or Utah, or West Texas—are ventifacts.

But wind-carved buttes all over Utah, for example, don't count as such. Nor slump-shouldered hills whose once mountainous brows wind has ever so slowly made off with. No, those are "deflations," more remainder than sculpture, despite the ruddy, troglodyte features that look down from their heights everywhere in Monument Valley, and fail to notice us. For just now, though, I stretch *ventifact* to include me, my hair, my nylon hood crackling in wind—though some gusts make me less "ventifact" than "deflation." That latter word is best reserved for phenomena such as Big Hollow, west of Laramie. There one can see absent kilometers and tonnages of soil long since removed by Wyoming winds single-minded enough to be in the business of excavating. If locals sometimes call them "blowouts," who'd argue? They're places wind remembers well, and works on.

But how are we to measure such weather? By the cubic yard of soil blasted into the air? We gauge, for example, morning calm by the way smoke rises straight up out of chimneys. If smoke drifts, but in currents so slight that weather vanes don't react, we say, "Hardly any wind at all." Later we notice breeze on the cheek; maybe weather vanes stir. Leaves and small branches toss lightly. Wind begins picking up dust, swirling loose papers around. Branches thrash. Strollers beside the lake see whitecaps cresting on its waters. The largest branches of big elms sway. Power lines swing back and forth. A woman opening her umbrella feels it go wild, try to yank free. Children walking to school stagger forward squinting, fighting not to lose balance. Whole trees start bucking and plunging like broncos. Limbs tear loose from cottonwood trees and go flinging downstream. Shingles fly through the air above rooftops. A parked car's rear window caves in. Stud walls of houses under construction buckle, as plywood takes to the air.

Each of these stages marks greater velocity, and can occur almost anywhere; in contrast, Wyoming people speak of making your own wind gauge, fit for Wyoming conditions: "You take about twelve foot of as heavy a log chain as a big man can drag, and a dozen or so railroad spikes for fixing it good to an oak stump. Now, if them links only stir a little, expect calm weather to continue. If your chain thrashes around some, there'll be a breeze by afternoon. If it should lift and stand straight out like a flag, the next day will likely be unsettled. But when your log chain gets to where it's whipping and popping and snapping off links, look out. That's a Wyoming storm warning."

Toward the microscopic end of wind's influence there's "saltation," tiny leaps and bounds made by sand granules as they're blown. Oh, I feel a bit of that too, when the wind's blast is loaded with snow crystals. Meanwhile, true saltation occurs constantly on the summits overhead, where sea beaches begin as a speck that ice, wind, and rain pry loose from its cliff and send spinning, tumbling. From there to Mississippi's or Alabama's Gulf Shore is but a question of time and spring runoff.

In southern Colorado, at the Great Sand Dunes National Monument, we've about forty square miles of sand that wind has piled up against a sort of cul-de-sac formed by the Sangre de Cristo mountains. The highest dunes rise seven hundred feet above the valley floor. Far from fossilized or coarse, they're shifty, fine-sifted, pure. Thus a park ranger once startled me by saying that when the National Park Service first built there, and sand was needed for mortar, it was trucked in from outside. I gaped, incredulous. Surely that was carrying the concept of eco-fragility to absurd limits. The dab of mortar used now and then for setting posts and other campground maintenance would hardly violate forty square miles. He was pulling my leg? But no. Those duned and wind-driven granules have endured so many thousands of years of saltation they've tumbled themselves round-shouldered, and thus lack angularity enough to bind really well in mortar. "So," that ranger explained, "when foundations needed pouring, better sand was trucked in."

By now I've reached the half-naked banner trees verging on Black Lake's east shore. Owing to wind they dare not branch out on their west side, so their lee-ward streaming boughs give an impression, even in calm, of speeding in place. I pass among them, and through, up a steep rise that keeps my skis pretty much out of sight, powder-whelmed; but the cliff I'm intent on comes into view: an ice-fall whose half acre plates of grey-blue runoff have wept down over tawny bulges of granite and frozen like armor, or like the ever-thickening shards of a beetle. By edging a few meters forward I cause the ice's gray-blues to tinge with five fathom green, or with zones of turquoise, subtle and eerie. The shifting colors fascinate, as if hinting at a world in which the whole spectrum could be frozen.

The granite wave they embellish so arrestingly is itself the size of intimidation. Megalithic, tremendous, it rises just three or four hundred feet above the white plateau of Black Lake. Gawking from its base, I feel the ice-and-stone presence of numb forces crushing my next thought so completely it becomes a barely inhab-ited stare.

I decide to boost energy and morale with a few hunks of milk chocolate. As has happened before, however, the cold makes it wholly flavorless. My mouth must be nearly wind temperature. Though I chew and chew till the squares feel like saw-dust, they don't melt or release the slightest taste. I tell myself, "Wait, it'll warm," and I do wait—a reasonable while—but to no effect, so I lose patience and swal-low. Chocolate incognito, I find, isn't chocolate.

Not having seen another soul for hours, my earlier self-interrogations now recur to mind; I realize that along with good old everyday animal fear, what I love about skiing alone in Glacier Gorge is the abyss between me and its presences. The more distant the nearest things are, the more dubiously I'm tangible. Or even here. Yet I know their "grandeur" is me. Has to be. I who am minuscule. Without my en-docrinal secretions, however, and the neural responses they trigger, all blue-black cliff shadows and white sheernesses remain a closed circuit. Poised on skis, find-ing no signs now of animals—human or otherwise—where my own tracks disap-pear, I scan panoramically while mulling that over. The feel is of being in the world alone. And not as if.

High in winter mountains our spontaneous awe makes a gift of itself to what caused it: snowfields whose whitenesses are absolute and untracked emotions—even if spattered with rock-and-ice Gothic. And light that can't make up its mind, shifting all day through cloud, and through the snow-shouldering, sun-starved forests. Gargoyle crags where silence and wind are never home the same time.

But ice-quarried rock, we know—or believe we know—is only the random effect of cold and crustal upheavals. Why should raw bigness summon the deepest, oldest feelings we life forms can have? Perhaps by the very size of indifference. Because mountains scorn the astonishments they give rise to, because they pretend to live entirely within the limits of the visible, because they despise our memories, we respect the hugeness of their refusal to confide. Which awes us. And we say so—inwardly. Meanwhile, snow and rock read our minds perfectly. To show their contempt for our least selfish thoughts, they ignore them. Whereupon we're awed all the more.

And we're grateful. Among fellow humans we're merely superfluous at best; at worst, part of the competition. But winter mountains enlarge the needle's eye of our tiny brains and their labyrinthine trivialities. Thanks to the rude unity of winter's fourteen-thousand-foot peaks, we feel our insignificance expand like a strange prestige—which makes being alive a kind of magic, easy as being not quite real. Small wonder that wherever terrain permits, primitives go around filling their habits with mountain gods. As for such earth spirits, we moderns—we whose houses are built of retired forests, we whose bodies are made of matter everlasting, and who each night before bed may drink a few sips of cloud—never think of them.

My fingers and toes remind me, "Get moving. " Yes. Often my pause turns daydream—till, finding myself at a standstill, leaning forward on ski poles, I waken when wind smacks the body heavily as cold water. Other times, as my heart pounds with me uphill, the hours of accelerated thump lull my senses into hearing neither its red hammer nor the thump and thud of weather. Suddenly wind stops: I panic, feeling my heart has too. Then wind resumes and so, its seems, does my high altitude heartbeat. This would be entirely ludicrous if the sudden scare were less chilling. Somehow my years in mountains have tricked me into confusing wind with life-signs in the world's body, including my own. That's nonsense, of course, but something in me doesn't know it. It's that something which gets ambushed by a sudden confusion, mistaking life and mountain wind for one thing.

Embarrassing, wholly unscientific. Yet where life forms are concerned, atmosphere on a windless planet might indeed feel inert, cadaverous. Columbine pollen would drop right back onto the lip of its bloom. Thistledown would fall at the base of each stem that contrived it; so maybe there'd be only one, a husky stalk clotted with dwarfish progeny. Pine forests, I recall, are among the earliest living exploiters of wind power—without which the gold smoke of their pollen would be a fine needle-sifted rain confined to the tree that released it. Groves just one tree wide. Happily, our actual forests wander like migrant tribes, mile after wind-sown mile.

As it is, spring pools often wear yellow films of wafting pollen; their stones are banded and waterlined with it. I've seen impulsive bursts of April wind huff pollen clouds for twenty, thirty yards still in the shape of its pine, or the shape of whole tree-clusters atop a mesa—the soul of a forest suddenly taking flight. Inland temperatures on a planet lacking convection would depend on solar rays, period. The sea would be glass. Sand wouldn't dune, and lacking saltation, mightn't even be sand. Snow wouldn't drift. Cities would be visible beyond the horizon as puddles of fume. Weather would be predictable as arithmetic, or yesterday. Nobody would run up a flag.

And I'm not sure all this snow would be here. Microscopically fine dust, as things stand, gets blown about the globe, providing nuclei for flakes and raindrops to crystallize around or condense on. And fallen snow often takes to the air again. I look up where Thatchtop Mountain's high ridgelines smoke with vortices of blown crystals that wisp and unravel like unruly root systems changing natures with fronds. Backed by an expanse of clear, blank-blue and half-violet high-country sky, snow's turbine shores feed back into air just now raking Mt. McHenry. Solar explosions prism within crystalline veils, and become sun dogs. Fast updrafts arrest themselves, then plunge, as if terrified of their own intentions. Powdered torrents sweep down like waterfalls, coil upward again like sidereal nebulae whose spiral arms pull themselves limb from limb, or ambitions whose first maneuvers destroy them.

Wind's motions: a mind always deciding, never made up. Why should its habits differ wholly from ours? Beyond the practical aims of food and shelter, we use motion to assure ourselves a future—as if movement kept our possibilities alive. Thus in watching Wagnerian winds rake the summit of Thatchtop, I feel the paradox of wanting to come again—to where I'm this minute standing. Wanting to see again what I'm seeing right now. A round-trip inside moments of standstill.

Skiing back downtrail I pass the Y-shaped writhe of a particular bristlecone pine I've known for years. If trees can have doting admirers, this one has me. Today, for no reason, I decide it's female, and christen her Yvonne.

As bristlecones go, her few hundreds of years make her a girl, but her figure nonetheless is centuries of violence enacted. I've only to glimpse Yvonne's evergreen torso and limbs to know what wind's sadism feels like. Even the slab her roots knuckle down into wears the wind's name etched into its granite like wood grain, a past still in progress.

All at once I'm ambushed by a snow devil, whose bluster can't stand contradiction. Its spindrift blinds me, makes skiing impossible. I halt, bobbing in it like a middleweight boxer, gasping for breath sucked from my lungs. Slowly my clenched eyes dare to blink open, wet, as the snow devil roars elsewhere. Through clotted lashes I watch its local shambles of powder kick up a ruckus as it goes. Across the gorge, high on the sides of Half Mountain, acres of granitic ruck smoke and howl. Three thousand feet overhead the flat summit of Longs Peak is also getting dismantled one sunlit grain at a time. Air's wildest routines, its iciest ways to catch fire include sudden geysering swirls torn off snowfields on quirks of velocity that spiral high, higher, then explode when cross-currents attack them, just to have something to kill.

Impossible not to remember Ruth Magnussen. And because I'm looking at the flank on Longs Peak where he came to grief, James Duffy III. He and his mountaineering buddy were trying to bag two peaks on the same outing. Though young, they were experienced; but their decision to go light, leaving heavier clothing out of their packs belied that experience. Coming down off the second peak, Longs, in colder wind than they'd bargained for, they began going into hypothermia. Duffy grew cranky, irrational. Should they try somehow to shelter among boulders and wait it out, or descend? They couldn't agree. By this time, Duffy's talk wasn't making good sense. The friend took off downtrail.

Desertion? To have stayed with Duffy up there in such weather might well have finished them both. If Duffy could hold out, the friend would bring help. But they had already spent many thousands of calories climbing a pair of difficult summits. And it's a long, rocky trail down. On such trails the last mile feels like five.

When at the Longs Peak Ranger Station the volunteer men and women of Rocky Mountain Rescue finally began uncrumpling from Jeeps, vans, pickups, and Volvos to head uptrail, in foul weather, hoping to find Duffy still alive, they were not mountaineer novices. Nor are Rocky Mountain Rescue people a gang of beer bellies in four-wheel drives. They are lean, fit, and—unfortunately—well-practiced. And very well equipped. Such persons take outdoor gear seriously. If you trek around much in Colorado's mountains you hear of people who didn't. James Duffy's story was even now being added to the local repertoire.

True, there's a strain of morbidity in rescue work. That day, despite the optimism of their effort, they must've guessed they were likely to find, if anything, a young corpse. What they guessed, however, I do not know. I do know that if "heroic" has meaning, men and women of Rocky Mountain Rescue have often been heroic. And if at thirteen thousand they report running into a snow-charged, eighty-mile blast that forced them back down the mountain, I *believe* in that wind.

But it lifted. Or enough. On reascending, what their search finally turned up was a young man in the prime of life, except for being dead. Deranged by severe hypothermia he'd apparently tumbled some 150 feet. What weather had begun, head injuries ended. Duffy was added to the list of those who've died on Longs, a list over forty names long.

Skiing over the scene of summer, now altered almost beyond recognizing, admiring wind-billowed snow as it streams thinly, beautifully off the highest ridges, admiring the late-afternoon blue of clear sky, I'm aware of watching snow blown around by nuclear fusion. If, as is true, solar energy drives Earth's weather, all winds, strong or gentle, arctic or tropic, arise at the sun's center.

Despite the fact that our planet's energies take millions of years to seep outward from the sun's core to its surface, the very cold that now tries to refreeze my left earlobe, once frostbitten severely, owes its windchill to that frantic solar core, whose fusioning produces temperatures at such astronomical degrees as to leave imagination helpless, empty. Looking around at a seeping cliff's icicle thickets, at creek ravines cargoed with drifts, I'm bemused by knowing that beneath Earth's crust, deep in its inner core, resides heat rivaling that of the sun's surface. Meanwhile its brilliance storms inside Thatchtop Mountain's veils of ripped snow. One cloud remnant opens, surprisingly like lids of an eye filled with the blank blue of the sky behind it.

As afternoon wanes, Glacier Gorge quickly chills. In open shadow its east-facing snow slopes go bluer, while sun on opposite slopes tinges warmer, more golden. My pole-baskets leave wells of aquamarine light, delicate, glacial. Many levels under their lovely, cold omen abide our oceanic progenitors: strata of tropical fossils, stone fish swimming within the currents of Lower Permian time; then Mississippian levels, then Devonian clamshells, brachiopods, and—for all I know—imbedded in the Cambrian strata, trilobites by the billions; whose light was the way we came.

About twenty-five miles under those life-cousins, Earth's crust gives way to the mantle, and that in turn to the so-called outer core, mostly iron in a fluid state. It has always bemused me that people who perish in local storms—such as I'm watching now—freeze while superheated winds of liquid iron stir at snail-slow velocities beneath them. I imagine white hotness, an incandescence that is at the same time thrice-fiery and red; yet black, totally. Its magnetic fields tell the lost traveler's compass needle which way to point. And that needle obeys, reliably, even after the mittened fingers holding the compass may have frozen solid.

What fascinates yet further is to realize that underneath Earth's superheated outer core lies an inner core so compressed that its superheated iron can't budge; is thus, at 7,200 degrees Fahrenheit, frozen. When gravity's avarice grows that intense, its familiar tug seems evil. If wind begins at the sun's center, Earth's inner core is surely wind's antipodes: a white-hot blackness, or so we imagine; yet red as lava. Planet dregs compressed to an anvil.

Though when wind rages the crazy flailing of spruce boughs seems less dance than anguish, maybe forests consider it exercise. Could it toughen their fibers? My reverencing of great-hearted fir and spruce trunks is partly rooted in these atmospheric devastations they defy, and thrive on. But wind is now upping the ante with so much blown powder I find myself skiing downtrail in whiteout. Visibility gone, I stop, wait. The blank air roars, a local blizzard that snuffs thought. A snowy nirvana.

The fact that *nirvana* combines Sanskrit words for "blow" and "out" strikes me as funny. But this cold air isn't fresh from India, and I'm not liberated, just erased; the weather of the mind is words, and for the moment wind has blown every word

out of my head. Asian wisdom might do the same thing, though I suspect any so-
journ in either Nirvana—as place—or in the fabled well-being of Eden would feel
to us like the Doldrums. Wind is itself a kind of emotion, even when it behaves
like a wild breathing no one has mastered. In windless Paradise there'd be no reason to stir, perfection being what and wherever you are, and each moment identical to the moment before. Instead, we postlapsarians are happily animate in a
fallen world—which compared to Eden has every disadvantage but one. Ours, at
least, never holds still.

That makes for a chaotic existence just now in Glacier Gorge. Would it seem
so, I wonder, if I watched by the eon instead of an afternoon? To stand in this
snow-smothered creekbed and stare unblinking, tireless as gravity itself through
a half-dozen centuries of wind, might shake my faith in chance. Snow's identical
patternings, the madly rhetorical gestures of fast moving clouds, even the warp in
wind-twisted trees—might prove regular as the zodiac.

I arrive at Mills Lake. Winter or summer, its windward shores are log-jammed
with grey creatures who, could they but talk, might settle the matter. But they're
not trees any more, just carcasses. When alive they met these very winds and won,
or stayed evergreen enough to believe they were winning, or believe—if they
weren't—winning was possible. Their twisted grain has long since been picked
into relief by "ideal particles" and high velocity snows. Icebound now, blank as
drained calendars, with the thaws of late May they'll revive their hopes of drowning themselves, of at last rotting away, dissolving downstream. They've been at it
for decades, and still wind isn't finished with them.

On one hand, the squalor of mere physics; on the other, something beyond
physics, something for which the lowliest snowshoe rabbit munching spruce tips
in the gorge is just one of innumerable masks. Though for long stretches this earth
goes mute on me, I still expect it to be a speaking world. And because we're made
of the same stuff it is, I wonder if certain of its voices—which we humans may
never evolve receptors for—aren't there, all the same, just below the threshold of
feeling, however inward and still.

Facets of our hidden natures may be less well concealed than we think. At times
their secrets seem to stream outward, as if to disguise themselves in this world's often alien promises. Looking up into air's gaudy ways with fast cloud—whose edge
sunlight iridesces to a hazed, transhifting rainbow—I half suppose I'm looking up
at one of my own interior descriptions. At the same time, I know only too well

that cloud-tatter is real, "other," apart; part of the sky's punctually restless mechanics. Of which I'm led by either culture or instinct to say, "Beautiful!" And I do.

But the fact we like it here, often immensely, wholeheartedly—was *that* inevitable? In wondering, "Why these things and not others?" I can easily imagine a rational species whose nature is simple as a two-spoked wheel: "GO/NO GO" or, "NIGHT/DAY," "USEFUL/USELESS," "YES/NO." The luck of imperfection, however, impels us exuberantly beyond so starved a response.

But exuberance won't get us past cold fact. "Face it," says fact; "Those voices you hear in wind aren't there." True, no matter how often I've heard dead trees cry. In thick forest that's a sound the wind makes: a dead trunk gets caught in its topple and is held, supported by living trunks which—for decades, as may happen—won't let that dead one fall, like our habit of taking along those who couldn't make it. When wind tosses the living tree crowns, that grey trunk rubs audibly against them, as if a creature dismayed. It's my voice that makes it a creature. Ours. The kind of voice we humans lend things. Especially whatever sighs as it ends.

Fact: the late afternoon turns colder. Time to go. When weather's calm, the very act of turning to leave seems sad; far less so when it's this windy. For the steepest few miles of descent I want ski boots firmly laced, and must take twelve or more miserable minutes to do what—without windchill—I could manage in 120 seconds. Removing pile mittens, I quickly unsnap my left gaiter, find its zipper-pull with numbing fingers, and unzip. Then I must thrust the bare hands into my knicker pockets to warm. I clench and unclench them, moaning a little. For their next sally forth, my fingers pick at and untie the yellow bootlace's firmly knotted bow. Back into the pockets they dive. Still cold but not frozen, they emerge to partly unlace and reface the boot, till each finger feels like a zombie—though oddly enough not the thumbs. Then more warming. Whereupon they must rezip and resnap the Goretex gaiter—hardest of all. After that I coddle them considerably longer, so that they can repeat the whole process on the other boot. They cower in my pockets like children begging to stay inside.

Before setting off again I look up at the sunset-tinted shambles of granite above me, and startle. As if I'd never been here before, as if I hadn't been glimpsing on and off for hours that long familiar hulk, now its mountain cragged with twelve-story spicules, its colosseum-sizes flakes, its rude chunks and smithereens seem wholly new; wanting to say something important. But it doesn't. Is itself, no more. Winter stone.

I turn from its crush and look back at the gorge headwall. Iron peaks. Against blue sky deepening nearly to indigo, loose snow spumes from their turrets in the frenzies and onslaughts I've watched all afternoon. Except that now, their distant, slow motion detonations—soundless as dust blowing off the moon—spell out a presence openly hidden.

The lowering sun reddens monolith knobs and snags and minor summits. It sets snow cornices afire till the westernmost ridgelines become one blazing rim. A ski pole in each mittened hand, I lean forward windburnt, chilled, transfixed. Veils of thrown crystal that plume over high desolations of granite, and over my own whiff of existence, become a "we." We who are. And are being annihilated inside processional volumes, each enclosing the other beyond comprehending. For once in my life I see all that is or can be, felt not as a thing but a power; and that power, a unanimous, self-radiant motion.

I see that we life forms stir, move to and fro a while, then slump down; even while wind grinds our cities to powder. Like everyone's, my birth and extinction always were, there, inside what I'm seeing. So.

Real as a gun muzzle, the eye of my death aims straight into me, and fires. My body is shot with its own blood; its redness is fear at the level of despair. Not one of my personal molecules will ever again take anything personally. The pure animal truth of being wiped utterly *out*. That terrifies me. Yet I feel the rightness in that, as part of what is. So be it. For once in my days, I *know* myself to be completely inside an implacable and luminous power, which, though itself, is also the power of a dark absolutely without edges. Its vastness infuses the delicate force of each particle of each atom. Each echoes it, just as each echoes all others. Tiny globules; pale, spherical, vibrant. Tiny intensities of blue fire, fire-flakes thick as snow. Each tiniest globular intensity hums with a humming *imparted* out of the very dark against which each is made visible; a darkness so encompassing, so total, it isn't there. Against which, all those invisible, particulate energies making up what we call the world's surface now pulse and float; one strange levitation.

I watch the skyline's western rim burn. In the backlit and prisming tatters of snow cloud, in their flash against the sky's high-altitude blue, I see that I'm to be destroyed more immeasurably than I'd ever dreamed of. The awareness fills me. My neck hair rises. This isn't "death," the thing people talk about, which is a mere mouth-sound, petty, inconsiderable. This is complete annihilation

But for the only time in my life I feel who and where I am. Truly within creation, no escaping. No place to dodge or hide. I was drawn forth. And am being even now scattered, taken back. Into everything. Into something that destroys all it makes. As it must. Otherwise it wouldn't be what it is.

So I can't wish it different. Literally can't want to. I can't both see it and want to do that. All I can be is dumb, scared, fascinated; a "yes."

I've never felt so invaded, never *been* anything this true. Its fading feels like one tremendous presence gone. By comparison, "reality" feels starved. Even its grandeurs, meager; poor, pinched, desolate things. The gorge returns. Rock and snow slip back to be where I am, among fellow ephemera. The grandly stupid stones of Glacier Gorge, their dusk-blue ponderosities. I sympathize with them, with the dear, wind-haggard fir, living or dead, with the wind-flustered ravens, the pine martens. And with myself. I see that each of us will have been the only one of his or her species. Our entire, ingenious planet is a single blown snowflake. Is an eye blink.

Yet as if admitting there's no place to go more real than here, no terrain that goes farther, even the mind's built-in critic relents: "Well, at least you came, looked around. You spent the energy. I'll give you that much."

The whole glacial trough has by now filled with open shadow. Every drifted ripple and snowfield is a skylit and empty blue. So are the snow-swagged pine boughs. Many thousands of feet higher up, the summit of Longs Peak warms itself at remnants of sundown. With alpen-glow fast losing color across wind-bitten escarpment, I watch the final rays go dull, watch the life of all afternoon slowly turning to stone, as the sun's red gold gets sucked into granite.

There Is No Freedom Spring Air

Tõnu Õnnepalu
(translated by Külliki Saks and Eric Dickens)

There is no freedom spring air;
glowing and surging around us
on this festival of the last melting snow
when the alder catkins
are heavy with yellow dust
and the earth is heavy with water
and the eye fell heavy with the beauty of light
when joy
comes and races
hands free down the hill
the lark falls through the air
when the sky
enfolds us into its perfect void
whose colors have no names
blue blue blue
a thousand times everything
with different meanings but
the dead grass and the earth
smell as for the first time in my life
and the pines
begin to rustle
when suddenly
an icy cold damp gust of wind from the sea
and the pinetops swaying in the sky
as if in the midst of a great sadness for there is no freedom

Tuula Närhinen, *Anemographic*. Light bulb on branches with wind

Holy Wind Before Emergence

James Kale McNeley

In setting forth conceptions of the nature of Wind and sanctions for their beliefs about its role in human life and behavior, bearers of the Navajo oral traditions characteristically recite appropriate passages of the creation story. This myth accounts for the beginnings of existence, the birth or creation of various Peoples in worlds that are conceived to have existed beneath Earth's surface, the emergence of these Peoples through those worlds to the surface of the Earth, and the subsequent creation of the present world on Earth's surface and of its specific inhabitants. Beliefs concerning the attributes of Wind and the nature of its relationships with other living beings are to be found in accounts of the underworlds. Sources may be found there for the conceptions that Wind has existed as a holy being from near the beginnings of the Navajo universe, being endowed with the power to give life and movement to other beings and possessed of knowledge which it conveyed to the Holy People.

Many versions of creation begin with descriptions of light which "misted up" from the horizons of the otherwise darkened underworlds. These mists of light, dawn in the east, sky blue in the south, evening twilight in the west and darkness in the north, arose from the aforementioned cardinal directions which were also marked by four mountains. The mists are said to have been the means of breath of the mountains associated with them, and within each of them, in turn, was an inner form "just like a real breathing human" (Haile n.d.:2). In this way the mists of light and the mountains of the cardinal directions are believed to have been living, breathing phenomena.

It is from the cardinal points so conceived that Winds came to the inhabitants of the underworlds to give life and power to them, as in the following account in which First Man and First Woman, who had been produced by Earth, were given strength. In this version, a single mist or cloud became the source of the Winds as well as of the diurnal cycle of light phenomena:

> While they were waiting for strength they saw a Cloud of Light in the east which kept rising and falling. . . . While they watched, it turned black, and from that blackness they saw the Black Wind coming. Then the Cloud of Light turned blue and they saw the Blue Wind coming to them, then it turned yellow and the Yellow Wind appeared, and then it turned white and the White Wind came. Finally the light showed all colors and the Many Colored Wind came. . . . This Cloud of Light also created the Rainbow of the Earth and as the light changes it creates *hayołkááł,* the White Early Dawn; *nahodeetl'izh,* the Blue Sky of noon and also the blue that comes after the dawn *nahootsoi,* the Yellow of Sunset, and *chahałheeł,* the Dark of Night. These were also made for the people of the Earth, and each of these is a holy time of the day from which comes certain powers (Haile 1933: 97–98).

The life-giving qualities of the Winds are indicated in the subsequent passage:

> When the Winds appeared and entered life they passed through the bodies of men and creatures and made the lines on the fingers, toes and heads of human beings, and on the bodies of the different animals. The Wind has given men and creatures strength ever since, for at the beginning they were shrunken and flabby until it inflated them, and the Wind was creation's first food, and put motion and change into nature giving life to everything, even to the mountains and water (Haile 1933:97–98).

By one account, Wind created by the mists of light not only gave life to all living things but, as Supreme Sacred Wind, became the Supreme Creator of the Navajo universe:

> The mists came together and laid on top of each other, like intercourse, and Supreme Sacred Wind was created. . . . Supreme Sacred Wind lived in light and black clouds or mists in space (Fishler 1953:9).

In this version, Supreme Sacred Wind, a being having the form of a person and knowing all that happens, created First Man, First Woman, and such Holy People as Talking God, Calling God, Coyote, and Black God. These Holy People, and the others he created, derived their special powers from him: "After the Supreme

Sacred Wind delegated His powers over to the gods, down below, He did not have power over their powers, only in part" (Fishler 1953:9–28).

Whether conceived of as a Supreme Creator or not, Wind made life possible in the underworlds not only by providing a means by which breathing could occur but also by providing guidance and protection to those who were created or born there. This conception is elaborated upon in an account which is quoted at length below in view of its unusually extensive description of the role of Wind as a guide or mentor. What is presented as a single narrative is actually a compilation of myth fragments elicited from the informant (JT) during several different interviews. As is customary, this history recounts the beginnings of the Earth and of existence, again associating Wind with phenomena of light:

> Wind existed first, as a person, and when the Earth began its existence Wind took care of it.
> We started existing where Darknesses, lying on one another, occurred. Here, the one that had lain on top became Dawn, whitening across. What used to be lying on one another back then, this is Wind. It (Wind) was Darkness. That is why when Darkness settles over you at night it breezes beautifully. It is this, it is a person, they say. From there when it dawns, when it dawns beautifully becoming white-streaked through the Dawn, it usually breezes. Wind exists beautifully, they say. Back there in the underworlds, this was a person it seems. (JT)

The Peoples of the underworlds migrated from place to place, initially lacking in plans or a sense of direction:

> In this way they lived back then and from there they moved to the place called Navajo Land. From there it is called One Word, it probably being spoken according to one word. There they went around like sheep. Like sheep, they did not talk. Their eyes governed their actions—if they are going to go *this* way, *that* way they run! Then *this* way! That is the way it was. There was no language, so they merely looked at one another. What would be used (the words that would be used in the future) to talk to each other there were none. To go over there (to direct one another) they said, "whoo, whoo," and they understood very well. In that way it was called Word One back then. From then they were Holy People, all four-legged animals being people. First Man and First Woman led them.

First Man, First Woman, Talking God, Calling God, these Holy People lived in this way. . . . "I will see for you," he (Wind) said. So, in this way Wind was obtained. Wind existed first and they encountered it as a person. Then, it seems, he continued: "Things will be known by me. With you it is not known where you will go. You do not know!" he said. "I know!" he said. "I know all about everything. I know about what is in this Earth and what is on it," he said. "I am Wind!" In this way back then he told about himself. There he became that which gives life. "Thank you, you accepted us," they said.

In this way, according to the way of the two lying on one another, they started off. That which was Darkness and also that which was Dawn became Winds. The one that was called in that way (Darkness?) will be called Dark Wind. When blueness appears at dawn, that will be called Blue Wind.

What was thinking, what was Wind, became words (language). "I will inform you," he said. "How is he going to tell us? He is Wind so how will it be?" "That will be known among ourselves—now I will just tell you!" he said again. Four times this happened, but he did not tell them. He did not say, "I will tell you" (Wind is here refusing to tell the people of the means by which he will inform them of things).

So they went forth from there. Yonder is what will be called "the Place of Separation." Through here a river is flowing. There, very awesome red mountains are located. The land is the same—spreading out beneath the mountains there it is red, too. The river, also, is reddish as it flows. "It is very beautiful here. We will live here from now on," it was said. Already then it (Wind) was in use: "You will move to here!" it was said to them through here—here, it seems, through the earfold where it cannot be seen. When we poke our fingers or stick something in here we touch it and are aware of it. It (Wind) is in here. From there it speaks to them. "Settle right here," is was said to them. "Right here!" (JT)

Here, the informant has related how Wind told things to the people, that is, through a corner of the ear which could not be seen. This, as was mentioned above, is what Wind had refused to tell the people themselves—*how* he would speak to them. The narrative continues

"It's really true what he said!" (they thought). From that time, whatever Wind spoke, it began happening in that way. Here they settled and having settled there suddenly it was spoken again from here (inside the ear): "You will make a leader," it was said to them. "The leader will be foremost among you. You will live under him, he standing for you. He will be your continuous leader and in that way you will live orderly," it was said to them.

Some women are very feared. These women are strong leaders, they act strongly, they speak out. There was this kind, called Woman Leader. "What will she say about what was said to us from here (in the ear)?" (they said). "Come here, listen!" they said to her, and they told her about it: "I was told that back there at the place called Darkness Crosses we came upon Wind Person and there he planned for us saying, 'Only according to my way will things be.' This is the situation now," it was said to her. So then: "That's right, it is true my children, they say that it's true. It was spoken to us like that, and within us it [Wind] is placed and to us it speaks. It (he) is our thinking."

So for that reason, things will happen according to his (Wind's) will. From that time what is called "leadership was formed" came into being: "What is called leadership will be placed; what is called speaking will be placed. Now it exists. Perhaps you know how to provide leadership? Perhaps you know words which you will use to make speeches? Again it was said, "Yes, we know!" "Alright, now it will be!" Way back there is when leadership was formed.

Here, leadership was again placed. The words by means of which speeches will be made and the words by means of which threats will be spoken were placed. From here (in his ear) he (Wind) was saying that by his will "You will proceed in this way so as not to make mistakes—not to proceed just according to your own mind, just as you want to do." They followed in this way. In this way leadership was formed there, Wind telling the one who is Talking God about it through here (his earfold). (JT)

Various Peoples went around aimlessly like sheep, reviewing and interpreting this account, lacking language and without plans until they came upon Wind in the form of a person who knew all and who told them that henceforth he would speak to them (through the corners of their ears). Wind then gave leadership to them and the words by which they would exercise this, the knowledge and the means to lead themselves. Even then, however, Wind gave them only a restricted governance over their lives so that a leader would not be able to do anything that he wanted to, "just according to one's mind." Rather, Wind made available the means of speaking particular words, including the threatening words needed by a leader.

The actual exercise of leadership in the underworlds, as distinguished from its establishment by Wind, was by chiefs who were appointed in one way or another to perform such functions as teaching the people how to farm and calling them together each day in order to tell them what to do. Even so, the conception by native historians of a continuing influence by Wind over these chiefs is evident. For example, informant JT related that Wind told the son-in-law of Woman Leader to get up and go out on a platform each day and tell the others what to do. Similarly, informant JD while recognizing that First Man and First Woman were leaders of many activities in the underworlds said that Wind advised them what they should do:

> The way we are telling things to each other, back then it (Wind) in the same way told them, "You should do this, you (two) should do this," being told thusly. So it is true. Back then it (Wind) told them. They used to live by it. (JT)

The guidance and sustenance provided by Wind to leaders in the underworlds is the prototype of Wind's relationship to the Holy People during the course of their emergence to Earth's surface and of its subsequent relationship to human beings, the Earth Surface People in the world created on Earth's surface following the emergence from the underworlds:

> Dark Wind, Blue Wind, Yellow Wind, White Wind, Glossy Wind, we will speak in accordance with their will—it happened (was said) like this. So, accordingly, one who speaks in our behalf has been present alongside us from back then. These same Winds speak for us and spoke for our late ancestors. Our leaders who speak for us held on to the ones that guide us. So back there they (Winds) told us that we live in accordance with their wishes. (JT)

By one account, it is even due to Wind's precedent that Earth Surface People later assumed responsibility for each other: "Now Winds will take care of us from the beginning of our lives. In accordance with that we will take care of each other from now on." (JT)

References

Haile, Berard. n.d. *Where People Moved Opposite* (Text Dictated by Curly Tó aheed-líinii of Chinle, Arizona). Manuscript 63–5 (English Translation) and 63–6 (Navajo Text). Flagstaff: Museum of Northern Arizona.

———. 1933. *Chiricahua Windway, Told by Slim Curly.* Manuscripts 63–17 (English translation) and 63–18 (Navajo text). Flagstaff: Museum of Northern Arizona.

Fishler, Stanley A. 1953. *In the Beginning: A Navaho Creation Myth. University of Utah Anthropological Papers,* No. 13. Salt Lake City: University of Utah.

The Aeolian Harp: An Allegorical Dream

F. H. Dalberg
(translated by Maynard Schwabe)

The mind loves to dwell on the fading images of time past, and awakens them like fond dreams from the womb of the past. I love above all your hours of solemn festivity, the happy days of youth, which I lived through in *Parthenope's* alluring pastures.

Once I sat beside *Possilippo's* rock cave on the seashore, it was one of those lovely evenings whose ecstasy can only be enjoyed under Italian skies. The sun had dived low beneath the billows of the ocean and gilded the purple hem of the western horizon; a glowing haze rested upon the still ocean: magical half-light took the place of blinding sunlight. Mountains and valleys were tinged with the loveliest colours, and the twilight threw a grey cloak over the landscape. To my left *Vesuvius* raised its fiery head. Spread out before it lay *Pompeii Portici* and the proud *Naples.* The sea was dotted with countless sails; in the distance one's eye alighted upon the grey foothills of *Capri;* to one side a moss-covered rack wall rose up, its dark thoroughfare leading to *Barjae's* and *Missenen's* enticing pastures was enjoying this sweet evening with sober pleasure when suddenly a magical whispering sound began to make itself heard, excited by the wind, growing stronger: soft at first, then increasing, growing weaker with a dying echo, and finally stirring mightily like the deep chords of an organ. I was unable to tell whether it was sylphs' or sirens' voices singing from the womb of the waves—until I came to notice an *Aeolian harp* lying in the open window of the neighbouring villa. It was the first one I had heard.

Delighted, I listened to the soft sound. Who tells me, I asked, from which galaxy these divine airs flow? Are they the voices of deceased friends, or the echoes of a former age? Is it *David's* harp mourning over *Jonathan*? Are they songs of lament of *Malvina* over *Oscar*? Or the last sighs of a loving mother for her suckling which death robs from her young breast, as the North wind does the hard buds of the spring rose? Rushing waves are now mingling with these aeolian tones, and the soft fading of the Compline bells.

The shimmering moon is rising, [and] the night becomes more and more festive. Ever softer becomes the whispering of the harp. The numbness of sleep overcame me as the *God of Slumbers* touched my eyelids [and] ushered dreams towards me as I breathed quietly—not fearful ones from *Hecates'* realm: sweet dreams, like those given birth by the pink flush of morning.

And look! The beautiful, lithe form of a goddess stands enchantingly before me!

"I am," said she, "the nymph of this rocky grotto! Mortal one, whatever you seek in this vision, that will I reveal to you, and lead you to its origin. But follow me silently, for you are treading on holy ground."

We flew through ethereal skies. First she showed me *Virgil's* sacred home, surrounded by evergreen laurels, and [then] we began handing to the sacred shades offerings of pure fruit. Rapidly the procession passed through cultivated fields along *Boya's* coasts in the terrifyingly arid valleys, whose gloomy entrance led to *Orkus* and *Pluto's* night domain.

Virgil's spirit and the god with the golden lyre *Mercury* protected us from the dangers of this realm of terror. In the dark chasm sank the fleeting shadows of the night; the suffocating sulphur fumes of this wasteland disappeared. Bright morning laughed all around us. Ambrose scents proclaimed the Elysian fields.

After we had enjoyed the divine view, surrounded by heavenly winds, my inner soul felt purified enough to hearken to lofty sayings from the world of spirits.

The nymph began: "In the blue and eternally serene realm of the air, pierced by no mortal eye, sways the island of clouds woven from pure morning dew, and the scent of flowers populated by ethereal spirits. No burning sun ray penetrates its mist surround: only the pale silver rays of the moon give it light. There the storms on earth fall silent, and eternal peace reigns, undisturbed even by the soft breath of the zephyr.

"Rocked in sweet dreams the souls of men move there, transported prematurely from this life, their wishes unfulfilled on this earth: innocent children and happy lovers, friends too soon parted, youths, men, old men who missed the goal of their existence.

"Fate—that solemn goddess—condemned them to this lonely isle. With sorrow they looked back on to the too soon departed earth, and with longing towards the moon of their home, where choirs of early-departed sister-souls awaited them with impatience until they had completed their years of purgation. They retained only the memory of the smells of that earthly Spring they had enjoyed. Unable to communicate with each other, veiled in fading fog, they sleep like butterflies in their dark chrysalis. Thoughts, desires, flutter around them like soft-winged morning dreams; their sighs of longing are enveloped in their depths, like shoots of tender flowers in the embryo bud."

The *Goddess of Harmony,* while in her wanderings through Creation, saw this island of clouds. The omnipotent *Zeus* [had] delivered her the charge of binding together in love and sympathy all beings created alike, so as to calm any dissension, and to keep the laws and elements in an ever-identical mood. He gave her music as a traveling companion, to calm agitated minds from their suffering and mourning by means of a magic sound. The gaze of the *Slumbering Ones* moved the goddess. She decided to lift her ban—for she alone was capable of averting their inexorable fate.

This solemn judge of worlds and times listened kindly to the sweet voice, inclining downwards the iron sceptre and waking the spirits from their dreams. Then there gathered upon the island an harmonious choir.

Like the wing-beat of young swans when they heave themselves up from the bank and fly through the air, such was the sweet melancholy of the songs. They penetrated to the valley of the shadows beneath the clouds, where forsaken friends gathered around them in mourning beside the burial mounds, yearning for them. The songs of mortal longing mingled with the murmur of the sublime spirits. Softly and solemnly they sounded, like the echo of distant bells, or the rustle of the evening wind through the forest tree-tops. Excited by the call of the goddess the sweet tones harmonized and blended with the heavenly choirs [music of the spheres] in a song of harmonious unison.

Polyhymnia listened happily to the divine song [and] then said to the spirits:

"Leave your twilight isle: this island is to be transferred into a finely-shaped harp and you yourselves into harmonious sounds destined to animate the divine strings. Sing sweet songs and hymns which are pleasing to gods and men."

The goddess hastened off to higher spheres. The divine harp swayed there in the fine-spun ether. A deep silence reigned about it. The spirits attempted to strike up another song but they lacked the strength to activate the strings: at each attempt they sank back into their slumber.

The *Goddess of Harmony* observed these unsuccessful efforts and took pity on the helpless spirits. She sent the fleeting gods of the winds to free their tender wings.

Aeolus appeared surrounded by a crowd of his children—the zephyrs and, zephyrettes. Light and fleeting, they flew up and down the resilient strings of the harp like dragon-flies around the tender filaments of the flowers. Their loving embraces became tones and there started up a delightful chorus of unheard-of magic song, one moment loud and vigorous like the wind on the open hearth, the next soft like a gentle murmur in the reeds.

The assembly of gods hearkened to these unaccustomed tones. They told the swift god of the air to delight them as they fell asleep or upon sweet morning awakening, with this lovely music. *Aeolus,* too, often sank to the ground at the sound of the divine harp lulling lovers to their dreams with magic sounds. But their song does not speak to everyone. It only rings a sympathetic note in the tenderhearted.

In multifarious keys and modulations the strings of the harp whisper elegaic sounds tender laments which entrance the souls of men and earth and which rise up to the stars.

And then the nymph: I had scarcely taken in her words when there stood at my side a delicate aeolian harp.

"Take hold of it" said the nymph "it will lead you to the temple of music. If its entrance is prohibited to mortal man then you should view it from afar. What prospect would then reveal itself? In the most pleasant valley of *Elysium* lies a lake whose waves whisper harmonious chords as the morning dew forms over the water and dissipates soft choirs of sweet flutes and harmonica notes stream out from its billows. They are carried on the light wings of the wind to the seat of the immortal and mankind.

"An island emerges from the bowels of toe lake, in the middle of which shadowed by dark palm trees, a sublime temple rises, resting upon crystal pillars, transparently polished and gleaming with the seven colours of the rainbow."

The nymph continued: "In the temple reigns the *Goddess of Time,* invisible to your earthly eye as she is limited neither by space nor earthly form; free without bonds and earthly fame her live spirit prevails ever giving birth to new life new transformations; lovingly she watches with maternal tenderness over the beloved daughters of *Harmony* and *Song* who rest in sisterly union upon her bosom. The choir of the friendly hours dance around the throne, and the genius of poetry, mime and rhythm—those winged messenger of time—envelope the heavenly host."

"But that lake," I asked, "whose waves splash around the temple?"

"That," interrupted the nymph, "is the *ocean of sound*. Through its midst glides the silver stream of song away on its peaceful course. It rises from its source in the dark cavern in those rocky heights you can see in the distance. It is called the grotto of resonance. Here all musical bodies are brought to life. Its cool interior hides the holy source of the song which soon splashes its way out in many varied keys and modulations—one moment like foaming mountain streams dythyrambs in powerful rhythms; then in a gentle tempo, gliding down long and soft on the melodic waves of the stream. Its flower banks are inhabited by the elementary spirits of sound, tender sylphite beings who live only in pure air.

"The souls of great musicians turn back after fulfilled earthly lives to this shore, they bathe in the waves of the stream and are purged in the higher secrets of their art in the crystal-clear water of its source. To you too, O son of melody, the entrance is not barred if the genius of music art considers you worthy of baptism for indeed I should not be guiding you any further since the red glow of morning already beckons me to leave this domain and return to my cavern. But approach the places of mystery. Sanctify this harp of the *Goddess of Resonance* who dwells in the rocks and the spirits who animate this harp will guide you further."

The nymph disappeared and with her the sweet dream. I awoke and there remained only the memory—*that I too was once in Elysium.*

Wind, Play with Me: The Mysterious Aeolian Harp

David Rothenberg

We have all tried in vain to capture the wind. It blasts against our faces, but we cannot see it. It pulls our hair back, but we can't wrestle it away. It screams through the air with a whistle and roar. It lifts off, and the sky returns to silence. The wind sounds like everything and nothing.

Can you play the wind? Can the wind play you? There was once a musical instrument played only by the wind, whose strings would resonate magically as the breeze grew stronger. Giant, round overtones cascading upon one another as the ears flowed through sound. Melodies created by vibration alone. Sounds so evocative that nearly all the major writers of the nineteenth century were swept away by their beauty and used the aeolian harp as metaphor for the way the earth could play with the soul. Coleridge, Shelley, Spenser, even Emerson and Melville were suitably impressed. Why was the figure of the aeolian harp so important for the Romantic age? We can't own the wind or easily convert it into music. It remains more.

You could say the aeolian harp is the string equivalent of the wind chime. Like most other musicians, I can't stand wind chimes. We don't like them because they strike the same burbling notes whenever the air blows. I already hear enough circling chords in my head. I don't need these pesky pentatonics joining in every time a breeze picks up. They're a cheap trick, an overplayed soundtrack for the universal breath. But the aeolian harp is a bit different. In its standard form, it's also not much to look at, just a rectangular box with four to twenty long, open strings, ready to resonate when held up to a gust. The sounds you get are stranger than the chime, because they are not tuned to any scale but the natural harmonic series, similar to the overtones you get if you blow across a bottle's mouth with a range of strengths, or the harmonics that come when you touch and release a single unfretted guitar string. Octave, fifth, fourth, then somewhere between a major and minor third, that soulful blue note, or shall we call it a "wind" note, that vibrates inside us so completely even though it's smack in between a piano's black and white keys. A pure property of sound is something our Western musical history has seen fit to dance around, to cloak, to adjust, to ignore like our own imperfect temperament. Our entire modern, tempered scales are slightly out of tune, the better to modulate between them. The pure resonances of the aeolian harp, like most of the world's untempered music, starts in one key and stays there for the duration of the sound, therefore always sounding more pure, revealing a bit of the lie behind our stratified musical civilization.

Bach knew all about this. His *Well-Tempered Clavier* was written to stress the differences in sound from one key to another; on today's modern instruments, all keys sound similar. The better to adjust to your range, my dear.

So whenever we hear pure, resounding overtones, sliding from one to another, it grounds us, speaking of something primeval, natural, something that existed long before, that will remain long after we've gone.

These are Romantic sentiments, for sure—the search for significance in a sound because it points out what's wrong with modern life. For people in the past, it was the idea of objects and systems that could work together and sound alike. For us, it is the dream of a natural world, the yearning for sounds that capture something pastoral, green, an airy utopia that never existed and probably never will. Ironically, not even the aeolian harp resonates nature. Play it for someone blindfolded today and they're liable to say it sounds like weird contemporary experimental music, or, if they fancy themselves in the know, they might say it's obviously some ambient electronica from around the turn of the twenty-first century. It is environmental, relaxing perhaps, but still eerie, strange, easy to hear as the result of machines, not the whipping of wind.

So is the aeolian harp a sound whose significance has slipped as fast as a pocket of dead air? We have the technology; we know how and why it works.

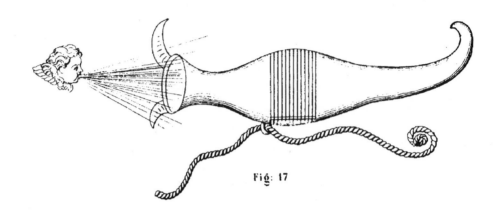

Fig: 17

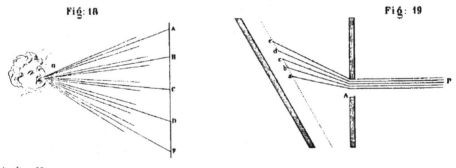

Fig: 18   Fig: 19

Aeolian Harps

But my main question still stands: Do we *hear* something fundamentally different from what the Romantics heard? Think of all the places music has gone in the past 150 years. Do we know more, or less, about the tones the wind can blow?

I suspect we hear something even less familiar, even more strange, but with greater context and more openness. There is more music easily available today than ever before, and more different camps championing all kinds of possible set-aside sound. With so much to choose to listen to, it is harder than ever to compare one kind of music to another, first of all in style, and second in terms of significance. How can we possibly decide which of the many sonic offerings actually matter? It's more than your likes and dislikes, but a question of where music comes from and what it can possibly become.

In order to understand the aeolian harp, we must start at its beginning. We can go back to the seventeenth century for a fine rendition of how to build one:

> Let the instrument be framed of that most vibrating wood, the linden tree, of which musical instruments are usually made, five spans long, two broad, and one deep; and then let it be furnished with fifteen strings of the same size, taken from the intestines of animals, or, there may be more, or fewer. The method of tuning them must not be by thirds, or seconds, but they ought to be all in actual unison, or in unison by octaves and concordant in one tone. The manner of application is this; when the wind blows strongest, let the stream of air be made to pass through a very narrow crevice of a window-shutter or door, to which crevice let the instrument be reclined, so that the wind must of necessity brush the strings. Then the master, without any further trouble, opens the window about half a span's breadth in the valves, in which the Aeolian instrument is inserted, and it produces such an unusual kind of harmony by the sole impulse of the wind, that no one (unless in on the secret) can conjecture from whence or what it is![1]

F. H. Dalberg, in his precarious fantasy reprinted in this book, heard "the strings of the harp whisper elegiac sounds, tender laments which entrance the souls of men and earth, and which rise up to the stars,"[2] all welling up from some misty "ocean of sound." Coleridge, in perhaps the best known poem on the instrument, "The Eolian Harp," finds it chock full of pregnant and penetrating metaphor:

> And that simplest Lute,
> Plac'd length-ways in the clasping casement, hark!
> How by the desultory breeze caress'd,
> Like some coy maid half-yielding to her lover, . . .
> . . . the long sequacious notes
> Over delicious surges sink and rise,
> Such a soft floating witchery of sound. . . .
> . . . Where Melodies round honey-dropping flowers,
> Nor pause, nor perch, hovering on untam'd wing ! . . .
> . . . Where the breeze warbles, and the mute still air
> Is Music slumbering on her instrument.

I've edited out all the harpless bits of this stanza to heighten the weight of the overtone imagery. For what can really be said of the power of pure sound? It's whatever you want to hear, isn't it? Can we call it a coincidence that the aeolian harp actually sounds not too far off from the singing of telegraph wires that later poets and adventurers were to praise? And heard from the twenty-first-century perspective, we might not think singing wires, but some phasing in and out on a synthesizer, in some glow-in-the-dark-blue nightclub room. Is it natural, or all the more mechanical? Is it still the sound of the wind that you think you hear now?

Stanza five of Shelley's "Ode to the West Wind" finds only ambiguous emotion in the surging overtones:

> Make me thy lyre, even as the forest is:
> What if my leaves are falling like its own!
> The tumult of thy mighty harmonies
>
> Will take from both a deep, autumnal tone,
> Sweet though in sadness. Be thou, Spirit fierce,
> My spirit! Be thou me, impetuous one!

No human art can know the indescribable sadness, the longing for a song of the earth that actually makes sense to us! The wind will carry us, our strings inside and out, those of the heart and of the tensions that cage us in, the human grid of the world, all reverberating sweetly in the air. So perfect that happiness and sadness are at last knowable as one!

This is an instrument that leads so many to genuinely complex feelings. Berlioz heard one and challenged us in 1844: "I defy you not to experience a deep feeling of sadness, of surrender, a vague and boundless yearning for another existence."[3] And Emerson had the harp itself speak up somewhat tritely in his "Maiden Speech of the Aeolian Harp":

> Give me to this atmosphere,—
> Where is the wind, my brother, where?
> Lift the sash, lay me within,
> Lend me your ears and I begin....

Melville's got one listening out by a stormy sea, with a title that sounds like a jazz record, in "The Aeolian Harp at the Surf Inn":

> List the harp in the window wailing
> Stirred by fitful gales from sea:
> Shrieking up in mad crescendo—
> Dying down in plaintive key!
> Listen: less a strain ideal
> Than Ariel's rendering of the Real.
> What that real is, let hint
> A picture stamped in memory's mint....

I never knew Melville went for poetry. Only in the wind, it seems. He goes on to describe a famous shipwreck and the breaking planks, the sunken reef, the lost crews no one will see again. A perfect storm. Indeed, what can you say about it but the noise and eternal power of the earth? What's the soundtrack for that movie out of memory? Swirling overtones, impossible to place. Nothing but the ambiguity of the famous aeolian harp, for its tones have always been more than sweet. A traveler describes a flying one he heard in Indonesia, the Asian variant—not a box on a windowsill but a wild singer perched on the tail of a kite:

> It was by no means a pure angelic nature which descended on us there. Demonic, wildly bewitched cries intermingled, a deep wailing tone, first falling, then turning into *the savage, terrible growl of a hunted, ferocious animal with no escape.*—Yes, who shall say what kind of unliberated spirit lay there behind the Milky Way was sending down to us the song of the raptures of his soul and the terrors of his heart?[4]

You see what rhapsodies simple overtones can invoke in the bewitched and bewildered listener? Indeed, so prevalent was aeolian harp imagery in its day, 150 years ago, that whenever people heard the wind in the trees, they almost automatically thought of the harp, even though so many more had read about the instrument than actually heard one. Here's German poet Otto Prechtler: "Hearken! From the mountain gorges comes an impetuous wind. A powerful noise stirs the air, like a funeral chorus. Wild desire pours out in heavy sobs, the spirit bursts from its bonds. I hear the sound like the cry of a swan inviting me to die. O wild whirlwind of song, take with you my soul. *Raise me far above this lowly earth!*"[5]

Curious yet? Admit it, you've never heard one yourself. Or maybe you've heard it and not realized what it was.

I first heard one through headphones in a tiny trekker's hut in the hills of Nepal. How exotic, you say. Actually not. A fellow traveler motioned me over, noticed my dreamy, faraway look. "This should take you closer to home," and he gestured to the collection of ten cassettes he had recorded to carry with him all over the world. This was 1982, in the days prior to compact disks and tiny digital chips that can contain more music than anyone could absorb in a lifetime. "Try this out." I put on the phones plugged into his first-generation Walkman. The music began with a deep primeval rumbling that swirled all around. Then saxophone and guitar; I smiled. Two of my favorite musicians, Jan Garbarek and Ralph Towner. Hyperboreal, reverbed romantic music, improvised, glorious, searching on. My kind of thing. I looked around the smoky room, a steaming cup of milky tea approaching my direction. The warm liquid calmed the throat. The band resounded on. Duet explorations blended with the sound of an aeolian harp left to tremble by the North Sea. The recording, *Dis*, which I would later learn meant "haze," a classic piece of Northern European improvised music. It may in fact have been this very experience of hearing this tape in the Himalayas that lead me to move to Norway four years later and spend many months there climbing, singing, and playing in the cool crisp North. The aeolian harp added gravity to this recording, quivering in sound like the silent curtains of the glowing auroras in the frozen winter sky.

See, the aeolian harp tends to do this to you—send the listener into an effusive tizzy, gushing on about the deep significance of sound. It's a classic response to a powerful drone, the same principle behind the grounding background pipes on a regal Scottish bagpipe, or an eternal raga coaxed to life on a sitar with a tambura beneath it—musical meaning that can have no beginning or end. The drone is almost more serious than the melody in these pieces, touching us deep at the center, joining our freedom of expression with the world.

It's so strange that the wind, fleeting and impossible to pin down, can conjure up these emotions. Elsewhere in this book, the art of Tuula Närhinen reveals the markings of the wind; she has tied pencil leads to tree branches and set them up to mark the trajectories of air. So this is what the wind looks like! Or is it those black-and-white arrows of air currents the weathermen use? The sound of the wind is either a gentle rustle in the trees, or a terrifying whistle swirling around the winter snow. Impossible to record with a microphone, because the machine perceives it only as noise, meant to be kept out by a "windscreen."

But the aeolian harp turns the rush of air into an everlasting thrum, or what the Zen masters call "a boom that fills up empty space." The sound hasn't changed for a few hundred years, but my question is: Do *we* hear it any differently now that music has become so effervescent and free?

I think we do. The key history that changed our understanding of swelling and singing aeolian string drones is the stringing of telegraph wires across the world's open spaces. These wires turned out to make a similar sound, now a byproduct of something practical, no longer a weird harmonic mystery. "Ah," the newly modern mid-nineteenth-century listener might say, "the dulcet tones of progress." Now our wires are either dissolved to the wireless or buried underground or else thickly insulated so that they thud instead of howl when tickled by the wind.

But we come back to the swirling tones of mystery in a musical climate that increasingly searches for pure sound. The twentieth century was the time of in-your-face nonconventional arts. Throw all the rules away! Make anything possible! Celebrate chance happenings and the jumble of all senses thrust together in accidental wonder! Once all rules have been broken, can anything still impress? Even if nothing is left to shock us, we still have to make sense of all this opportunity. Just because anything can be art doesn't mean everything *is* art. It must be praised, held up, framed, listened to with greater attention then ever before. This will be the century of looking, listening, sensing. It's not going to be easy.

For a long time they've been telling us that any sound can be music: a scratch, a moan, a tree falling, and the rush of wind on a pure open string. But we who are accustomed to hearing almost anything coming out of calm, dark stereo speakers have come to like our sounds removed from context. If you hear a recording of an aeolian harp, you'll swear it's something experimental, a symphony conjured up digitally inside an obedient machine. Wordsworth had no record collection; his music had to come from something tangible for it to exist at all. No wonder he and his cadre were so amazed.

They heard a pure sound from the future, a glimmer of the abstract beauty that would call the whole idea of authenticity into ultimate question. The aeolian harp cannot be placed today, seeming equally natural and artificial at the same time. In our whole world of anonymous noises, it is gentle, inoffensive, easy to be lost.

But if you pay attention, you will catch a bit of purity that can never be denied, despite what I said—something timeless reverberates through the ages. You can go back to the pre-Buddhist Bon religion of Tibet to find an eloquent paean to this perfection:

> The Way of Pure Sound is the way of change.
> Without avoiding, it seeks to accept.
> Taking all into friendship, everything is its friend.
> With all as your friend, nothing stands alone.
> Sky and space, method and wisdom;
> losing duality, reaching consummation—
> Perform a realm of perfect enjoyment.[6]

The swirl of sheer overtones remains long after the wind has died down. This logic of sound is part of the structure of the world. Once you hear it, you'll know what the air sounds like. It is the harmonic answer to the white-noise whoosh of dry fall leaves.

I only hope you're still able to hear it someday, out there in the real world with a strong breeze on your face, a sound that won't disappoint. Music can't be easily described, and I can't be the one to tell you what you will think when you hear the aeolian harp. The rise of industrial machines may have numbed our reaction to these strange drones, but the fact that the wind can play music on wires hung high is yet one more reason to respect the air in all its force and desire. There is no end to the chance for songs to descend upon us down from the sky.

Notes

1. Stephen Bonner, "The History and Organology of the Aeolian Harp," in Stephen Bonner, ed., *The Aeolian Harp* (Cambridge, U.K.: Bois de Bologne, 1970), 2:74.

2. Andrew Brown, "The Aeolian Harp in European Literature, 1591–1892," in Bonner, *Aeolian Harp,* 3:280.

3. Ibid., p. 73.

4. Bonner, "History and Organology of the Aeolian Harp," 2:790.

5. Brown, "Aeolian Harp," 3:850.

6. David Snellgrove, *The Nine Ways of Bon* (Berkeley: Dharma Publications, 1970) 12.

The Dimmitt tornado. Courtesy of the NOAA Collection

The Fallacy of Safe Space

David Keller

August 11, 1999, started out as a normal errand-study day. I had left my car with a mechanic, who was kind enough to lend me his, and headed to get some coffee. Tired from a long bike ride the day before, I was content to sit outside a café, sip a triple iced espresso, and look at people strolling in the sun. En route, I had listened to the local radio news. According to the National Weather Service, warm, moist air was moving in from the south, with colder, drier air coming in from the northwest. Afternoon thunderstorms with possible rain and hail were predicated, typical of late summer in northern Utah.

Gazing became procrastinating, and I turned my attention to thinking about strategies for revamping the environmental ethics course I was scheduled to teach autumn semester at Utah Valley State College. I needed to overcome a problem I had the first time I taught the course: the equivocation of the study of "environmental ethics" with "environmentalism" in the minds of the students. I teach in a very conservative community, and environmentalists are considered subversive to traditional values and a threat to free markets. In a previous course, I thought I would gain favor by bringing in a guest scholar to give a lecture titled "Mormon Sources for Environmental Ethics." To my horror, as the lecture ended, a student approached the speaker, red-faced and angry, exclaiming, "The Church authorities have not issued any statement about this! To imply that Latter-Day Saints should be environmentalists or that Mormon doctrine implies environmentalism is disobedient!" Remembering that moment made me shudder.

This time, I thought, the solution was to explain that environmental ethics deal with human choices about the environment—choices that might be entirely oriented toward human ends. On this approach, one might hold that humans are disconnected or separate from nonhuman nature in some fundamental way and that the use of nature for human benefit is morally justified. Since developing an environmental ethic in a semester paper would be a major requirement of the course, it was important for me to make it clear that defending a human-centered position was entirely feasible.

To emphasize the idea that the Western tradition has been characterized by a pervasive anthropocentrism and that in one sense the study of environmental ethics is simply an investigation of this aspect of Occidental culture, it occurred to me that a brief historical account was appropriate. The Genesis creation myth in which God said, "'Let us make man in our image, after our likeness'" and the injunction, "Let them have dominion over the fish of the sea, and over the birds of the air, and over the cattle, and over all the earth, and over every creeping thing that creeps upon the earth," would provide a starting point recognizable to the students.

I planned to move along history by pointing out that ancient Greeks shared the Hebrews' human-centered perspective. The *polis,* the political manifestation of rationality, separated Athenians from the chaos and barbarism beyond the city wall. Socrates, in fact, claimed that he had nothing to learn from the trees and open country outside the city. The idea of a human-nature divide reigned twenty centuries later when the French philosopher René Descartes asserted that his identity as a human being—the possession of an immaterial, eternal soul—had absolutely nothing to do with having a body. The views of John Locke would also strike a familiar tone to the students' ears, since Utah politicians repeat over and over again a Lockean mantra when arguing against wilderness preservation: nature has no value until humans use it and therefore should not be "locked away."

I paused and took a sip of coffee. While these ideas would no doubt resonate with most of the students' own views, I needed to draw these threads together with a tangible, contemporary illustration. I recalled how I flew to Miami from Quito, Ecuador, shortly after Hurricane Andrew and was startled by the degree to which modern buildings can take a pounding: the sturdy, well-built homes of the affluent along the Atlantic coast remained intact, while entire neighborhoods of flimsy homes farther inland were demolished. I could use this illustration to point out that technology, in the form of building innovations, has furthered the notion of a separation between human beings and the rest of nature that extends all the way back to antiquity. Concrete, steel, and glass can withstand forces far greater than the mud and timber structures of the past. Few people wanted to live on the Florida panhandle or in the desert Southwest before the invention of air-conditioning. Improved building technology has successfully distanced us from nature's capriciousness.

The sun disappeared behind a surprisingly dark cloud, and I took off my sunglasses. The formerly gentle breezes began blowing my notes off the table and flipping the pages of my textbook forward or backward twenty at a time. I closed the book and placed the glass coffee mug on a pile of papers.

Even if I succeeded in making the point that the study of environmental ethics was well within the purview of Western Christian culture, I still needed to make it apparent that the anthropocentric worldview is not the only alternative. Prior to Judeo-Christian monotheism and Greek rationalism, our nomadic forebears probably did not see themselves as apart from nature. They were always *in* nature. There was no wild or wilderness from which they shrank. With the advent of agriculture, linear furrows and regular inundations must have generated a sharp contrast from famine and flood. In a sense, humans became distanced from the rough natural vicissitudes with the advent of a sedentary lifestyle. A constellation of historical factors—agricultural, political, philosophical, religious, and so forth—converged to form the conception of human separateness from nature, a conception that was questioned by the greatest minds of our time.

At this point in the course, a student would invariably ask me about *my* stance on the issue. The notion of a fundamental human-nature divide, I would have to respond, was in my view a serious fallacy. But I would need to defend this statement in the light of the numerous technological advances I just pointed out that seem to distance us from the capriciousness of natural process. I wasn't quite sure what I would say.

Looking up, I noticed another thunderhead forming over the valley, growing gradually taller and darker. The underside of the cloud was black and muscular. The pouch-like texture was at once beautiful and ominous, yet there was no lightning or rain. The shifting breezes did not abate, and I decided to head up to the University of Utah library for the rest of the afternoon. I gathered up my materials, stuffed them into my briefcase and then headed back to the Audi borrowed from my mechanic.

Halfway to the University of Utah, I noticed a swirling cloud body in the shape of an inverted cone dropping from the black mamma above, directly over downtown Salt Lake City. The northwestern breeze coming off the Great Salt Lake must have come into contact with the southerly wind. The convergence of these two surface air masses was creating a weakly rotating system.

The cloud looked like an inverted cone boiling with turbulence, swirling noticeably in a counterclockwise direction. It was an incredible sight, and I drove up the hillside to get a better look. As I drove, the cone became part of a full-fledged funnel cloud extending from the valley floor to the thunderhead. An enormous plume of swirling garbage was sucked into the sky. All the foam, cardboard, newspaper, and small leafy branches seemed suspended, circling slowly above the city. It was one of the most awesome and beautiful things I had ever seen. It looked like a tornado, although I hesitated to assume it was anything more than an enormous dust devil, since Utah is not known as a tornado state. As far as I could tell, the funnel appeared to be stationary. I figured that if I saw it moving toward me, I'd simply drive away.

In the few minutes it took me to reach the top of a residential area northeast of downtown known as the Upper Avenues, which has commanding views of the Salt Lake Valley, it had grown completely dark. Debris rained out of the sky. I turned into a dirt pump station parking lot that joggers and walkers use to access the open mountainside above the houses. Blowing dirt erased any view beyond the edge of the parking lot. The car was engulfed in a sandstorm. A violent blast came directly at the windshield as I watched the debris go from paper to two-by-fours. A large piece of plywood flew straight at me and then spun suddenly off to the left, like a giant Frisbee. Aware that somehow I was suddenly in the funnel, I shifted the car into reverse in an attempt to escape.

I had backed up only a few feet when an electrical wire fell across the roof of the car, accompanied by a loud crash. I stopped. Two feet behind the car lay a splintered timber power pole. A transformer hit the ground, spilling out fluorescent green fluid. Relieved that the power pole hadn't crushed the car, but afraid that the wire on the car was hot, I quickly pulled forward, and the wire fell to the ground. The surge of wind quickly waned as hail began to fall. The violent gust had lasted for only ten or fifteen seconds. After five minutes, the hail slackened, and I got out of the car to look around.

The power pole I had almost backed under had been snapped off about ten feet above its base. Nearby, two other power poles had been downed. Criss-crossed wires covered the ground. A portion of a roof had landed on the bare hillside. The trunks of dozens of trees were snapped, and others were uprooted and plowed over. Debris was scattered everywhere. Lawn chairs, shingles, a charcoal grill, a swamp cooler were just some of the random items decorating the parking lot. Part of a roof hung in a pine tree. Half of one house was gone. Tangled venetian blinds dangled in shattered windows. A large wood beam punctured a huge glass window and black smoke rose above downtown.

As I surveyed the aftermath, I was struck by the incredible folly of my actions. I assumed that when I saw the funnel cloud developing over Salt Lake City, I could observe it in its full sublimity from a safe distance—like the confident observer in David Casper Friedrich's painting *The Wanderer*.

People started to come out and look around. It was dead calm until sirens began to wail. A man who was inside his house when the tornado hit said the sound was deafening. Strangely, I didn't remember any sound.

An employee of the local water utility emerged from a windowless brick pump station with a look of bewilderment. He called the central office on his two-way radio, and we learned that, according to preliminary police reports, the tornado developed southwest of downtown, traveling northeast, skirting Temple Square (for Utahans, an example of God's will or sheer coincidence, depending on perspective), and advancing up the hillside to my viewing spot. The water utility employee reported that the pump station was inaccessible due to the downed power pole and requested that the power company be dispatched to remove the obstruction.

As I surveyed the mayhem the tornado wreaked on this carefully manicured and ordered upper-class subdivision, it occurred to me that our success in increasing the amount of physical space that we can manipulate has promoted a false sense of security, a hypertrophic confidence in our inventions. This actual physical space corresponds to the socially constructed ideal of safe distance or safe space. The concept of safe space is the basis of distinguishing the human from nature, a central feature of the Western worldview.

Yet not all of our beliefs fit reality. Safe physical space, the tangible, technological product of rationality, is conflated into the *concept* of safe space. Once this concept works its way deep into our cultural psyche, we are poised to posit a fundamental boundary between humanity and rest of nature. As the tornado demonstrated, there is no such boundary. The human-nonhuman divide is fallacious, and I am guilty of this erroneous conflation.

I stared at the roofless house. Through a front window, I could see kitchen cabinets and the refrigerator. The open sky above cast a gray light throughout the rooms, not normal for the inside of a house. Its roof lay at least a hundred yards away, to the west. The direction of the blast that ripped the roof from the house moved east to west, but the direction of the blast that came at the car moved west to east. I had been inside the vortex. The "safe distance" I thought I had—several miles—suddenly became the distance between my face and the windshield. The car protected me from the blast, but the threshold between the glass shattering or not must have been precariously thin. If the car had been perpendicular to the blast rather than nose first, the window might not have held.

After several minutes, I became aware of voices around me. Somebody said that one of us should go into the house to look for survivors. Somebody else said the family was out of town. Another suggested organizing cleanup crews. I headed back to the pump station parking lot to see if the downed power pole had been removed, so I could return my mechanic's car.

As I walked, weaving a serpentine path through broken branches and scattered shingles, I admitted to myself that for good or ill, inventing and making defines humanness. Technology is intrinsic to being human. Constructing houses, building dams, playing musical instruments, practicing medicine, and catching flight are integral aspects of the human identity. Obviously, we cannot forswear technology, but we need to recognize that one thing humans cannot make is an inviolable safe space. At best, we may temporarily distance ourselves from the wild forces of nature.

Near the pump station, I came across an utterly shredded copy of the *Shvetashvatara Upanishad,* a peculiar find considering the overwhelming probability of finding the *Book of Mormon* or *Doctrine and Covenants* instead. Along with many other ancient Indian texts, this scripture describes the Hindu god Shiva's dance as perpetually sustaining the cosmos through simultaneous destruction and creation. As I recalled the image of the beautiful and awesome plume rotating over the city, with newspaper, dust, and leafy branches fluttering in a Dionysian frenzy, it occurred to me that the tornado was like Shiva—or *was* Shiva. The same ecological processes that drive evolution by natural selection also cause extinction. Like Shiva's dance, this dual effect of natural process is inextricably interconnected. They are two manifestations of the same underlying phenomenon.

Back at the pumping station parking lot, workers were cutting the power pole into five-foot sections and piling them aside. As I got into the car and turned the ignition key, it occurred to me that even I, who have devoted considerable time and energy reflecting on the human—more-than-human dynamic, fell into the trap of committing the safe space fallacy—almost fatally.

Acknowledgments

Thanks to William Alder of the National Weather Service for explaining the unique physics of this tornado and to my auto mechanic, Jim Keller (no relation), for not being angry when I told him I had driven his Audi Quattro into the vortex of a tornado.

Wind and Breath

D. L. Pughe

A week ago, Bellagio was still flush with foreign visitors wandering through the narrow, carefully swept cobbled passageways in search of scarves and ties or dangling wooden Pinocchios. They marveled at the perfect order of the small town nestled in the crook of Lake Como's two waving arms. Everywhere they looked were stained ochre walls and dark orange–tiled roofs, the whisked stoops of every shop and home, and animated citizens chatting in the central square. They sat in the sun lakeside or under an awning at the Lido or the Hotel du Lac sipping cappuccino or indulging in a decorated glacé. Now, a week later, the season is over as they say, and a cold wintry *Tivano,* or northern valley wind, has arrived. Huge scarlet leaves from the ivy that clings to the walls of Villa Serbelloni above have tumbled down into town, mixing with smaller golden leaves from all the trees along the way. The wind has come up, and the tourists are gone, and the stores are all shuttered with piles of leaves drifting along the passageways. Down the narrow stairs, into the square with its ancient chapel, and out across the vast pewter lake, the breeze gains momentum and increases its chill. Small whirlwinds grab up the leaves and spin them in the air like furious dervishes that tornado off, floating out over the whitecaps toward the opposite shore.

A week ago, the ducks were well fed from french fries and ice cream cones dropped along the lake's edge and were nonchalant about coming over to inspect offerings of crackers or bread. Now they scurry across the waves crashing against the gray sea walls toward anything edible tossed to them, with an anxious glance over their shoulder at the approaching winter. And the tall diagonal concrete stairs of the Lido's diving platform, stairs once ascending in sunlight toward some mythic heaven, are now stark against the leaden sky as though the rest of their steps had been broken off by the wind and washed away.

When the wind arrives in Como at the south end of the lake, it is even fiercer and roars down the streets of the city, forcing its citizens to huddle together in doorways. During malevolent gusts, *"Il tempo brutto!"* is murmured consolingly as a greeting. Only the most stalwart grandmothers can be seen with bundled children in the abandoned leaf-strewn playground of the central park near the wharf. There stands the Volta Museum with its impressive nineteenth-century dome, and yet it appears torn apart by the wind. It is only in a state of renovation but with chunks of marble and heavy equipment lying idle as though too cold to touch. Inside, the antique mahogany and glass display cases are shoved together, and only a few exhibits remain on view. The ancient batteries and glass bulbs, the brass rods and dark wooden implements, all forged during the earliest years of electricity, lie scattered some distance from the numbers that once marked their spot in each case, and they are all covered with a fine layer of dust. The attendants who run the museum, two elderly ladies whose days are spent playing cards with woolen sweaters tucked around them, seem surprised a visitor will still come by, much less pay to enter. They shrug, interrupted from their game, and take the money all the same, offering a thin guide to all that one cannot see but only imagine from the rubble both inside and out.

Later, out in the streets, young high school students are attempting a political demonstration about the recent horrors in our world. With a few home-made banners, a rag-tag group mobilizes in the city square, and many of the kids, their mufflers wrapped around ears and mouths, gradually defect due to the ferocious push of the wind. The ones who stay appear to be involved in adolescent flirtations with one another, with much joking and friendly shoves. Their earnest leader with the bullhorn has trouble getting everyone in order, and the police shiver and lean against their car, amused and unhurried. Finally, the charge is called, and one protestor ignites a red emergency flare that sends out vibrant crimson sparks; the wind takes the sparks up like the spray of a fountain. A small puppy along for the march is both attracted and frightened by the light, pulling forward and back, and the group moves off with much laughing and pushing, into the enveloping charcoal dusk.

When a cold, piercing wind howls into town, it seems a harsh injustice, a punishment for things gone wrong. In this autumn of 2001, the wind in Italy rings with all that has changed everywhere since a new kind of terror overtook our lives. Winds are only part of the complex currents of air that chase around the world. Unlike seas separated by land, the atmosphere is continuous, connected, far and near, global and local. A network communicating vast distances, back and forth, the wind is a messenger of both good and ill.

Names given to winds are *in statu nascendi,* born on the tongue in the heat of the moment. Here in Bellagio there is the *Breva,* a warm breeze from the south; the *Vento,* another cold valley wind; and the *Tivano,* the recent fierce gale from the north. And not far away in the Besgell Valley of Switzerland is the *Brüscha,* and in the pathway of the Rhine the robust *Wisper.* Toward the west in Spain is the *Criador,* a traveling disturbance that precedes rain, and the *Levanter* that blows in stormy gusts through the strait of Gibraltar.

The worst winds are notorious and are blamed for a rise in malevolence of human actions, for a malaise of the soul. On the Mediterranean coast, for example, a shrill siren will quickly rouse sleepy sunbathers to hurriedly grab their striped beach chairs and umbrellas, their towels and books, clamoring for shelter. Swimmers caught in the waves glance over their shoulders, then rapidly stroke to shore. Before the pulsing blare of the horn dies away, the fist of the *Sirocco* hits the dancing surf, and as it punches the beach, all the dust and heat it has carried from the dunes of the Sahara is unfurled. Then it grabs up the sand underfoot and whirls it all together in tormented gusts. Any poor soul still attempting to escape has no way to turn from the wind; it whips in all directions at once. Braudel tells us that as it passes through Saharan villages on its way north, it tears out gardens and orchards, reducing a year's work in a minute of destruction. Then, as though fortified by a hearty meal, it heads out across the vast Mediterranean full of great gulps of sand. When it reaches the famous beaches of northern Italy and the South of France, it throws itself about with a triumphant force and has been blamed for increased murders, accidents, anger, and suicide. In earlier eras, it was once possible to offer the *Sirocco* as an alibi for violent crimes, the culprit in spectacular cases of sudden, unconscionable homicide.

From the northern polar regions comes the *Mistral,* the "masterful" wind that plunges south through the Rhone valley of France in violent cold and dry gusts. Damaging crops, threatening the railways and challenging sanity along the way, it arrives in Marseilles, where it stays sometimes for a hundred days. It can hurl children into canals and pull off chimneys in ways natives call "malevolent" and "impetuous."

And tearing down from the Alps, the warm dry *Föhn* is a menacing and notorious wind. In southern Germany, Munich in particular, it is surprising at first—a warm gust during the thick chill of winter. But as it violently sprints down the mountainsides, melting everything in its path, torrents overflow the rivers; buds and blossoms mistake it for spring, only to freeze again and perish. It shakes the branches of trees like truant Katzenjammer kids caught by the schoolmaster, and it shrieks around the windowpanes, pounding on the glass. The *Föhn* syndrome includes anguishing bodily pains as well as headaches, dizziness, nausea, and fatigue. And it is blamed for every unbalanced psychological state: hostility, irritation, anxiety, depression, inability to concentrate, and occasionally elation and suspiciously excessive niceness.

The *Föhn,* too, has been used to defend the worst of human actions: "It was the wind that picked up the knife, stuck it in again and again." From afar, from places without wicked winds, it is always confounding when someone invokes the weather to explain away his or her guilt. But when you are caught in a wind, pinned flat against a building by its fierce intensity, hounded down the street, pummeled as you attempt a hill; after you have listened to it squeal around you day and night, night and day, for weeks on end; when it has kicked sand in your face and whirled back to do it again, it could possibly change your mind or even possibly help you lose it.

In Afghanistan and Iran, there is a wind of 120 days called the *Bad-I-Sad-O-Bistroz.* Careening violently downslope from the northwest, from the direction of Europe, it usually arrives between May and September. It blows continuously, and in the dry, furrowed dusty terrain of Afghanistan, it increases the hostility of an environment already unwelcoming to humankind. The world is alert to the complex drought-afflicted landscape of Afghanistan, and, with overdue humanity, is worrying over the plight of the people there. There are rumors the network of terror is housed in the caves sunk in the rugged Hindu Kush mountains. *Aeolus,* the ruler of winds, was himself a cave dweller, keeping his potent breezes tied up in a sack. It was curiosity rather than revenge, however, that led his powers to be loosed on the world.

We are just beginning to understand how the generous citizens of Afghanistan refrain from judgment as they welcome and care for visitors. Journalists, health workers, and *mujahideen* recount how Afghanis with little or nothing themselves will take you in, giving you their last drop of tea and their only bowl of rice. It is not the wind there that one can blame for all that has changed with the world.

To be in Italy in the fall of 2001 is to be strangely apart from our American plight and the Afghani perils, and yet it is possible to feel caught and pulled between the poles of Afghanistan and Manhattan. We feel the fierce winds of change, and yet the last image of New York, passing through on October 1 on the way to Italy, was windless, as though the breath were knocked out of it.

From friends who were in the city in the sizzling days of early September, we heard about an eerie calm, absent of even a *Cat's-Paw*, the most gentle of American winds. Perhaps it was most like what Leopardi experienced, where "not a breath of wind stirs a single leaf / Or a single blade of grass, and you can't / See or hear, near or far, a ripple of water / Nor a cricket chirping, nor a wingbeat / Flittering in leaves, nor an insect buzzing, / Nor any sound or any movement at all. / A profound hush settles." Perhaps it was like the last innocence before the storm.

A few days later, on a morning of absolutely beautiful hot clear skies and an absence of wind, a horror beyond imagination changed all our lives. We watched one tower blazing with flames and hopeless souls escaping the smoke, abandoning themselves to the air, falling countless stories in final agonizing moments of consciousness. Then another tower was pierced by another plane. The scene replayed itself until both towers fell, and a deathly stillness settled in the smoke and debris across a center of the world. The horror was indescribable, and when we arrived en route to Italy, we were drawn to the gaping wound of the city, covering our faces to blunt the mingled scents of disaster that hung over Lower Manhattan. We felt the absence, the towers that were never architecturally loved (though the people were) but how, as the highest beacon, they supplied Manhattan with a compass for figuring out where you were. As we gazed at the ruins, the fugitive limb of their presence and the agony of the families left behind ached at the back of our hearts.

The city was groping for air and at the same time letting go of any former feeling of safety and comfort. We recalled every haunting image, particularly a young man in a light jacket, his arms calmly at his sides in a headfirst fall. We realized the abject hopelessness that led him to cast himself into the windless air.

A *wind rose* is a diagram where the length of lines drawn to the center of a compass shows the frequency and speed of the wind. Where there is no wind, it appears to be crushed on one side. In the stillness of New York, it seems to embody a sense of direction gone awry, where part of the world has fallen away and a new angle has taken hold. It looks illogical from all that we know, neither reasonable nor understandable; horror arrived and left us all gasping in its wake. It was only later, in the relative safety of Italian shores, that I found consolation again in words, the first clues for how to weather the future. Centuries earlier, Petrarch had begun to fathom grief, to reassure us how, at such moments,

> Love, wisdom, valor, pity, pain,
> Made better harmony with weeping
> Than any other likely to be heard in the world.
>
> And the air and the wind were so filled with this deep music
> No single leaf moved on its still branch.

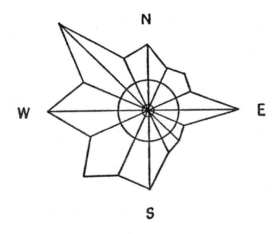

Wind Rose

Contributors

Manuel Acevedo is a photographer whose work has appeared in PS1's "Greater New York" exhibit and Bard College's "Re-Locations." He has worked with El Museo del Barrio and International Center for Photography combining drawing and photography in workshops for elementary school students in East Harlem. In 1992, he received the New Jersey State Council Distinguished Artist Award.

Stuart Allen is a photographer and sculptor whose artwork is found in many public and private collections including the Tokyo Kite Museum, the Crocker Art Museum, and the permanent collection of the U.S. State Department. He lives in Woodland, California.

Lori Anderson is the author of *Cultivating Excess* and *Walking the Dead*. Her newest collection of poems, *Persona,* explores the intersections of culture and nature through three personas: Canoehead, Bog Girl, and Subway Bride. She has an M.F.A. in poetry from the Iowa Writers' Workshop and a doctor of arts from the University at Albany.

David Appelbaum, professor of philosophy at the State University of New York at New Paltz, is editor of *Parabola Magazine* and publisher of Codhill Books. His forthcoming book, *The Shock of Love,* will be published by Lindisfarne Books.

Hayden Carruth, born in 1921, has published twenty-nine books, chiefly of poetry but also a novel, four books of criticism, and two anthologies. His most recent books are *Reluctantly: Autobiographical Essays, Collected Shorter Poems, 1946–1991* (awarded the National Book Critics' Circle Award), and *Scrambled Eggs and Whiskey* (1996), which won the National Book Award for Poetry. He has been poetry editor of *Harper's* and has received fellowships from the National Endowment for the Arts and the Lannan Foundation.

paulo da costa was born in Luanda, Angola, raised in Portugal, and has lived in Canada since 1989. He is the editor of *Filling Station,* a Canadian literary magazine. His first collection of short fiction, *The Scent of a Lie,* was published in 2002.

Marsha Cottrell has been awarded fellowships from the John Simon Guggenheim Memorial Foundation (2001) and the New York Foundation for the Arts (1999). In 2001 she was awarded a printing residency from the Institute for Electronic Arts at Alfred University, funded in part by the New York State Council on the Arts. The raw material of her drawings are keyboard characters that she organizes using a computer.

Friedrich Hugo Dalberg (1760–1812) was a writer, composer, and treasurer of Worms, where he was in charge of training musicians in the cathedral. He belonged to the circle of great early Romantic thinkers that included Herder, Rousseau, and Goethe. The most recent edition of his collected writings was published in German in Trier in 1997.

Susan Derges has established an international reputation over the past fifteen years with one-person exhibitions in London, Cambridge, Edinburgh, New York, San Francisco, and Tokyo. Her artistic practice has involved cameraless, lens-based, digital, and reinvented photographic processes as well as video.

Carl Djerassi, professor of chemistry at Stanford University, has received numerous awards, including the National Medal of Science for the first synthesis of a steroid oral contraceptive. He is the author of five novels, a collection of short stories, a poetry chapbook, and, most recently, the memoir *This Man's Pill* (2001). In addition to *Oxygen,* he is the author of *An Immaculate Misconception,* which has been performed throughout Europe.

Harold Fromm taught in the English departments at the University of Wisconsin, Brooklyn College, Indiana University Northwest, and the University of Illinois in Chicago. He has published a large number of essays and reviews in the *Georgia Review, Critical Inquiry, New Literary History*, and regularly for fifteen years in the *Hudson Review.* His books are *Bernard Shaw and the Theater in the Nineties* and *Academic Capitalism and Literary Value.* He is coeditor of *The Ecocriticism Reader:*

Landmarks in Literary Ecology. He lives in Tucson and is an ongoing visiting scholar at the University of Arizona.

Kristjana Gunnars is a writer and professor of writing at the University of Alberta. Her latest book of poems, *Silence of the Country,* was released in the spring of 2002. She has also written *Exiles Among You* and the prose work *Night Train to Nykøbing.* Her work has received a number of awards, including the Stephan G. Stephansson Award for poetry and the McNally Robinson Award for fiction.

Werner Herzog knew from the age of fourteen that he would become a filmmaker. He never attended film school but has made many of the late twentieth-century's most renowned German films, including *Aguirre: The Wrath of God, Nosferatu, Fitzcarraldo,* and *Where the Green Ants Dream.* He lives in Munich and insists that "film is not the art of scholars, but of illiterates." *Of Walking in Ice* is his only published literary work, and it has never before appeared in the United States.

Roald Hoffmann is a writer who makes his living as a theoretical chemist. He and Kenichi Fukui received the Nobel Prize in 1981 for their theory concerning the course of chemical reactions. Based on quantum mechanics, their theories discuss how atoms behave. Hoffmann writes nonfiction, essays, poetry, and plays. His latest books are his third poetry collection, *Memory Effects,* published in 1999, and *Old Wine, New Flasks: Reflections on Science and Jewish Tradition* (with Shira Leibowitz Schmidt), published in 1997.

Harold Humes, novelist and educator (1926–1992), cofounded the *Paris Review* literary magazine and wrote two acclaimed novels: *Underground City* (1958) and *Men Die* (1960).

Franck André Jamme has published ten volumes of poetry and fragments since 1981. He has also published numerous limited editions illustrated by artists. In 1983, René Char asked him to direct the publication of his complete works for Pléiade. As an independent curator, he has focused on contemporary Indian Tantric, tribal, and outsider art. Translations of his poetry are included in the anthology *The New French Poetry* (1996) and *The Recitation of Forgetting,* translated by John Ashberry (2002).

David Keller is director of the Center for the Study of Ethics and assistant professor of Philosophy at Utah Valley State College. His first book, *The Philosophy of Ecology: From Science to Synthesis,* coauthored with ecologist Frank Golley, was published in 2000.

David Lukas is a naturalist living in the Sierra Nevada foothills. His writing has appeared in *Audubon, Orion, Sunset,* and numerous other magazines. His most recent books are *Wild Birds of California* and a forthcoming revision of *Sierra Nevada Natural History.*

Nathaniel Mackey is the author of three books of poetry: *Eroding Witness, School of Udhra,* and *Whatsaid Serif.* He is also the author of an ongoing prose composition, *From a Broken Bottle Traces of Perfume Still Emanate,* of which three volumes have been published: *Bedouin Hornbook, Djbot Baghostus's Run,* and *Atet A. D.* He coedited (with Art Lange) the anthology *Moment's Notice: Jazz in Poetry and Prose,* and is professor of literature at the University of California, Santa Cruz.

Howard Mansfield has written about preservation, architecture, and American history for the *New York Times, DoubleTake, American Heritage,* the *Washington Post,* and other publications. He has explored issues of preservation in three of his previous books: *The Same Ax, Twice: Restoration and Renewal in a Throwaway Age* (2000), *In the Memory House* (1993), and *Cosmopolis* (1990).

James Kale McNeley served for many years as a social caseworker for the City of New York Department of Welfare, New Mexico Department of Public Welfare, and Bureau of Indian Affairs. He eventually married a Navajo teacher, Grace McNeley, and moved to Honolulu, Hawaii, where he received a Ph.D. in anthropology. His dissertation, "The Navajo Theory of Live and Behavior," was the basis for *Holy Wind in Navajo Philosophy.* McNeley is vice president of Navajo Community College/Diné College.

Edie Meidav wrote *The Far Field: A Novel of Ceylon* and is completing a novel set in southern France. She teaches in the M.F.A. program in writing and consciousness at New College of California in San Francisco.

Sarah Menin studied architecture at Newcastle University, England, where she holds a lectureship. She coauthored *Nature and Space: Aalto and Le Corbusier* with Flora Samuel and is writing *Relating the Past: Aalto and Sibelius.*

Steve Miles has taught writing and literature at Colorado State University and the University of Northern Colorado. He lives in Denver with his wife and three children. Other work has appeared in the *Chattahoochee Review,* the *Sun, Poem, Colorado Review,* and the *William and Mary Review* as well as *Seven Hundred Kisses: A Yellow Silk Book of Erotic Writing,* and *Fathers: A Collection of Poems.*

Arno Rafael Minkkinen is a Finnish-American photographer who has been working with nude self-portraits in the landscape since 1971. Once an assistant professor of photography at MIT, his monograph, *Waterline* (1994) was awarded Grand Prix du Livre at the twenty-fifth Rencontres d'Arles. His most recent book, *Body Land,* was published in 1999. Minkkinen is now professor of art at the University of Massachusetts in Lowell.

Bruce F. Murphy is a poet and essayist whose work has appeared in *Critical Inquiry, Poetry, Publisher's Weekly,* the *Paris Review, Pequod,* and other journals. He won the Bobst Award for his first collection of poems, *Sing, Sing, Sing,* and served as editor for the fourth edition of *Benét's Reader's Encyclopedia.*

Tuula Närhinen, born in 1967, works and lives in Helsinki, Finland. She records nature by various graphic means, capturing the movements of trees and water in the wind, trying to imagine the landscape as seen from the perspective of a bird, a fish, or a moose. Närhinen was the recipient of the Cité Internationale des Arts residence in Paris and a grant from the Finnish Cultural Foundation.

John P. O'Grady is the author of *Pilgrims to the Wild* and *Grave Goods: Essays of a Peculiar Nature.* He is working on a new book, titled *Occult Ecology: Reading Nature Darkly.*

Andrea Olsen is professor of dance and chair of the Department of Theatre, Dance, and Film-Video at Middlebury College. She is author of *Bodystories: A Guide to Experiential Anatomy* and a forthcoming book, *Body and Earth: An Experiential Guide.* She performs and teaches internationally and is currently touring a three-year performance project, *Path,* with music composed by Mike Vargas.

John Olson is the author of *Echo Regime,* a collection of poetry. His literary criticism has appeared in *The American Book Review, Rain Taxi, Sulfur,* and *First Intensity.* Olson is working on a book about air.

Jaanika Peerna works in photography, drawing, and digital imaging. She has shown in New York and in her native Estonia. She is completing an M.F.A. in visual research at the State University of New York at New Paltz and is a teaching artist with the DIA Foundation in Beacon, New York.

Stephen Petroff is a painter-writer who lives in Sagadahoc County, Maine. His imaginary or one-of-a-kind books include *Night: What Is It?, Crickets the Toolmakers,* and *Old Charlatan's Manual.* Books with broader distribution include *Red Tea*

Weather Report, Spirit of the Stone Thrown to the Bottom of a Lake, and *Head of a Bull on the Soundbox of a Harp.*

D. L. Pughe's essay on Leonardo da Vinci, "The Lost Notebook of Aqueous Perspective," was published in *Writing on Water* (2001). Other recent essays have been published in *Searchlight: Consciousness at the Millenium, The New Earth Reader, When Pain Strikes,* and *Nest* magazine.

C. L. Rawlins lives a wind-blown life in Wyoming. A field hydrologist, he won the U.S. Forest Service National Primitive Skills Award for scientific fieldwork in the Wind River Range. He has written two nonfiction books, *Sky's Witness: A Year in the Wind River Range* and *Broken Country: Mountains and Memory,* and two books of poetry: *A Ceremony on Bare Ground* and *In Gravity National Park.*

Reg Saner's essays and poems have appeared in more than 140 literary magazines, such as the *Atlantic,* the *Paris Review,* and the *Georgia Review,* and in over three dozen anthologies, including *Best American Essays.* His book *So This Is the Map* was selected by Derek Walcott for the National Poetry Series. Saner's backpacking, back-country skiing, and canyoneering have furnished materials for his *The Four-Cornered Falcon: Essays on the Interior West* and *The Natural Scene,* nominated for the John Burroughs Medal in nature writing.

Andrew Schelling teaches poetry, Sanskrit, and wilderness writing at Naropa University in Boulder. He has traveled through South Asia studying Indian culture and poetry, and has had work translated into German, Dutch, Spanish, and French. Recent books include *The Cane Groves of Narmada River: Erotic Poems from Old India* and *Tea Shack Interior: New and Selected Poetry.*

Ellen Scott is an independent artist living in New York. Her work features organic forms and behaviors in both process and content, ranging from fine art to three-dimensional animation, digital video, and interactive installations. She is finishing an M.F.A. in computer graphics and interactive media at Pratt Institute.

Virgil Suárez was born in Havana, Cuba, in 1962. He is the author of over twenty books of prose and poetry, most recently the collections *Palm Crows* and *Banyan.* He lives in Miami.

Louise Weinberg has had eight solo exhibitions of her work and participated in over 150 invitational and juried group exhibitions. She has won numerous awards for her photography and encaustic paintings, and her work is included in several books.

Crystal Woodward has exhibited her art and photography in France, the United States, and Germany. Her work has appeared in *Leonardo, The World and I,* and elsewhere.

                                                      ✒

David Rothenberg is associate professor of philosophy at the New Jersey Institute of Technology and the founder of *Terra Nova,* the annual book series on culture and nature. His books include *Sudden Music, Blue Cliff Record,* and *Always the Mountains.* Rothenberg is also a composer and jazz clarinetist, with five CDs to his name, among them *Bangalore Wild* and *Before the War.*

Wandee J. Pryor is the managing editor of *Terra Nova* projects at the New Jersey Institute of Technology. She received her B.A. from Colorado College, and prior to *Terra Nova* she worked in advertising for Random House and St. Martin's. She is the author of two plays, *And God Plays Dice* and *Rats Live on No Evil Star,* which have both been performed in New York.

Sources

David Appelbaum, "The Laugh" from *Everyday Spirits,* NY: The State University of New York Press, 1993. Reprinted by permission of the author.

Hayden Carruth, "Notes on Emphysema," Reprinted by permission from *The Hudson Review,* Vol. LIII, No. 4, (Winter 2001). Copyright © 2001 by Hayden Carruth.

F. H. Dalberg, "The Aeolian Harp: An Allegorical Dream," [1801] trans. Maynard Schwabe from *The Aeolian Harp in European Literature, 1591–1892,* ed. Andrew Brown. Cambridge: Bois de Boulogne, 1970.

Carl Djerrasi and Roald Hoffmann, "Scene 14," from *Oxygen.* This represents a revised version of an earlier published scene in *The Kenyon Review* (Vol XXIII Num. 2) and in *Oxygen* (Wiley/VCH, Weinheim 2001). Printed by permission of the authors.

Harold Fromm, "Air and Being: The Psychedelics of Pollution," reprinted from *The Massachusetts Review,* © 1984 by The Massachusetts Review, Inc.

Kristjana Gunnars excerpts from *The Substance of Forgetting,* Calgary: Red Deer Press, 1992. Reprinted by permission of the author.

Werner Herzog, "Of Walking in Ice," originally published as *Vom Gehen Im Eis* © 1979 Carl Hanser Verlag München Wien. Reprinted by permission of Carl Hanser Verlag.

Harold L. Humes, excerpts from an interview in Cambridge, MA, on December 13, 1981, part of an upcoming biographical documentary about Harold L. Humes. For information, visit www.doctank.com. Printed by permission of Harold Humes estate.

Franck André Jamme, "The Mystery of the Hills" from *Korwa Drawings.* Previously published by the Galerie du Jour Agnes B. and The Drawing Center. Reprinted courtesy of the author.